▶   The Moscow Pythagoreans

DOI: 10.1057/9781137338280

Other Palgrave Pivot titles

DOI: 10.1057/9781137338280

palgrave▸pivot

The Moscow Pythagoreans: Mathematics, Mysticism, and Anti-Semitism in Russian Symbolism

Ilona Svetlikova

palgrave
macmillan

DOI: 10.1057/9781137338280

First published in 2013 by
PALGRAVE MACMILLAN®
in the United States—a division of St. Martin's Press LLC,
175 Fifth Avenue, New York, NY 10010.

Where this book is distributed in the UK, Europe and the rest of the world, this is by Palgrave Macmillan, a division of Macmillan Publishers Limited, registered in England, company number 785998, of Houndmills, Basingstoke, Hampshire RG21 6XS.

Palgrave Macmillan is the global academic imprint of the above companies and has companies and representatives throughout the world.

Palgrave® and Macmillan® are registered trademarks in the United States, the United Kingdom, Europe and other countries.

ISBN: 978–1–137–33829–7 EPUB
ISBN: 978–1–137–33828–0 PDF
ISBN: 978–1–137–33827–3 Hardback

Library of Congress Cataloging-in-Publication Data is available from the Library of Congress.

A catalogue record of the book is available from the British Library.

First edition: 2013

www.palgrave.com/pivot

DOI: 10.1057/9781137338280

What was Pythagoras? <…> the Hyperborean Apollo.

(Iam. VP 140)

▶

… a race of mortals so singular in their shapes,
habits, and countenances.

J. Swift, *Gulliver's Travels*

DOI: 10.1057/9781137338280

Contents

DOI: 10.1057/9781137338280

DOI: 10.1057/9781137338280

List of Illustrations

DOI: 10.1057/9781137338280

Acknowledgments

It is with much gratitude that I acknowledge the assistance of the Russian Institute of Art History (Zubovskiĭ) throughout my research, and a grant from the FMSH (Paris) which supported a part of this study.

My thanks to my friends and colleagues Arkadiĭ Blĭumbaum, Milad Doueihi, Nikita Eliseev, Sergeĭ ÎArov, Irina Ivanova, Georgiĭ Levinton, Irina Lukka, Nataliîa Mazur, Maurice Olender, Stas Savitskiĭ, Patrick Seriot, and Sergeĭ Zenkin who helped me in countless ways.

The Department of Rare Books and Manuscripts of the Scientific Library of the Moscow University kindly gave permission to reproduce materials from its collection.

I am deeply grateful to Graham Stack for being such a fine editor, and to Eugene Ostashevsky for his efforts on behalf of this book.

If there is anything of value in my work, it is the result of unfailing help and encouragement of Omry Ronen. This book is dedicated to his radiant memory.

List of Abbreviations

GARF	Gosudarstvennyĭ arkhiv Rossiĭskoĭ Federatsii
IMI	Istoriko-matematicheskie issledovaniĩa
MS	Matematicheskiĭ sbornik
MT	Mirnyĭ trud
ORK i R NB MGU	Otdel redkikh knig i rukopiseĭ Nauchnoĭ biblioteki Moskovskogo gosudarstvennogo universiteta im. M.V. Lomonosova
OZ	Otechestvennye zapiski
RGADA	Rossiĭskiĭ gosudarstvennyĭ arkhiv drevnikh aktov
RGALI	Rossiĭskiĭ gosudarstvennyĭ arkhiv literatury i iskusstva
RGB	Rossiĭskaĩa gosudarstvennaĩa biblioteka
VFiP	Voprosy filosofii i psikhologii

DOI: 10.1057/9781137338280

Introduction: Belyï's *Petersburg* and Moscow Mathematicians

Abstract: *The political doctrines of the Moscow Mathematical Society, which have never been adequately investigated, offer rich material both for the history of modern ideology, and for the history of Russian culture (in relation to Andreï Belyï's seminal works). The present study is part of the history of Russian science in its relation to political life, as well as a contribution to our understanding of one of the greatest Russian writers.*

Svetlikova, Ilona. *The Moscow Pythagoreans: Mathematics, Mysticism, and Anti-Semitism in Russian Symbolism.* New York: Palgrave Macmillan, 2013. DOI: 10.1057/9781137338280.

In the nineteenth century the traditional Christian hostility to the Jews evolved to become a network of scientific and historical theories providing a purportedly solid basis for political and ideological speculations. Jewish influence became the object of ever increasing scrutiny, with its signs apparently ubiquitous. They were compiled and summarized by Houston Stewart Chamberlain (1855–1927) in his *Foundations of the Nineteenth Century* (1899) in which the course of world history is viewed as the product of a struggle between the noble Aryans and vicious Semites. This mystical conception of history recruited many believers. Sinister proofs of the Jews prevailing over the Aryans were found in political developments, as well as in the state of arts, sciences, and education. What follows pertains to the history of this mindset.

1

The subject of this book is the union of mysticism and politics in Russian intellectual ideology at the turn of the twentieth century, but the thrust of research is two-pronged: it addresses controversial chapters in the history of Russian mathematics as well as Russian literature through the analysis of a connected problem in one of the most abstruse masterpieces of early-twentieth-century Russian prose.

The action of Andreï Belyï's (1880–1934) novel *Petersburg* (1913) takes place during the first Russian revolution and its aftermath: terrorists, with a diabolical Jew as the principal actor, plot to assassinate a conservative senator; the senator's son gets involved in the plot.[1] Throughout the novel we are repeatedly reminded of the senator's passion for geometry. Most importantly, his thoughts of the ideal state are infused with geometrical imagery.[2]

Much of Belyï's novel is autobiographical. The main inspiration behind the senator was Belyï's father, the mathematician Nikolaï Vasil'evich Bugaev (1837–1903), dean of the physical-mathematical faculty of Moscow University and president of the Moscow Mathematical Society. The link between his profession and the senator's geometrical meditations must have seemed too obvious to deserve attention, and the meaning of this link was thus never explored.

That it should be explored, however, is immediately apparent from an article about Bugaev (1927) written by the mathematician Veniamin

DOI: 10.1057/9781137338280

Fedorovich Kagan (1869–1953) for the first edition of the *Great Soviet Encyclopedia*, volume 7. The following passage catches one's eye:

> <Bugaev> together with his disciples, Prof. P.A. Nekrasov being the most active among them, created in Moscow a whole philosophical school of strongly metaphysical character that had a considerable influence not only among mathematicians, but also in the wider circles of Moscow scientists. Some of the members of the "school" established a link between these philosophical views and the political ones of a glaringly reactionary quality [îarko reaktsionnogo svoĭstva].[3]

Even Marxist encyclopedias do not usually include any information about political views of mathematicians. The very fact that Kagan found it necessary to mention politics, and reactionary politics at that, in connection with Bugaev, suggests that the latter's transformation into a conservative senator in *Petersburg* had more to it than Belyĭ's mere whim.[4] Just as Ableukhov's love for geometry is not comprehensible without taking into account the profession of Belyĭ's father, the senator's conservatism is linked to the same profession. A commentary on the geometrical passage of *Petersburg* requires the study of an ideological trend which emerged among Moscow mathematicians: the attempt to provide a mathematical foundation for ultra-conservative doctrines, of which anti-Semitism was an integral part.

2

The ideology in question has not completely died out. Its traces are still discernible among contemporary Moscow mathematicians, a fact that indicates the historical significance of our subject.

Testimony to this continuity is, for instance, an apologetic volume on the life and activities of the publisher of the *Protocols of the Elders of Zion*, Sergeĭ Nilus, printed in 1995 by one of the most prolific authors on the history of the Moscow Mathematical Society, himself a graduate of the Faculty of Mechanics and Mathematics of Moscow University, Sergeĭ Polovinkin.[5] The main focus of his studies on the Society are the philosophical, religious, and political views of its members.[6] It will be noted that some of their works testify to an attentive reading of the *Protocols*, belong to the same ideological frame of reference, and are impossible to understand without taking this into account. Interestingly, Polovinkin, who, given his knowledge of Nilus, can hardly be unaware of this, never

DOI: 10.1057/9781137338280

alludes to the *Protocols* in his writings on the Society, nor does he ever speak of anti-Semitism in those writings.

This apology of Nilus by a devoted historian of Moscow mathematics provides us with a direct link between the ideological concerns of the Moscow mathematicians of the past and those of the present.

The political writings of another graduate of the Faculty of Mechanics and Mathematics of Moscow University, the eminent mathematician Igor Shafarevich, made him known far beyond professional mathematical circles. His anti-Semitic essay *Russophobia* (published in 1989) provoked a scandal that culminated in the US National Academy of Sciences asking him in 1992 to resign his membership. The conceptual core of *Russophobia* is constituted by a theory of two conflicting peoples within one state, a "small" one and a "big" one, the former undermining traditional values of the latter and posing a threat to the stability of the state. Shafarevich referred to the historian Augustin Cochin (1876–1916) who formulated this theory in his analysis of the French revolution. Cochin's "small people" consisted of members of the "societies of thought [les societés de pensée]," notably the Freemasons, advertising liberal values in the period preceding the 1789. While containing some interesting and original features,[7] Cochin's analysis of the French revolution was at the same time rooted in the reactionary ideology of his time. Whatever the distance between Cochin, who was a serious historian, and the fabricators of the *Protocols of the Elders of Zion*, they share ideological concerns: the importance attached to Freemasonry and, more generally, to the spread of "abstract" liberal ideas presented as responsible for the horrors of revolutionary destruction and the subsequent victory of depersonalized democratic regimes.

Prompted by Rightist beliefs and, most likely, by Rightist vocabulary as well, it was not difficult for Shafarevich to render Cochin's scheme anti-Semitic by identifying the latter's "small people" (or "small state," another of Cochin's expressions which in this context inevitably reminds one of the "state within the state" accusation repeatedly emerging in the anti-Semitic tracts) with the Jews.[8]

Shafarevich denied the allegations of anti-Semitism. The remarkable aversion of some anti-Semites to the word "anti-Semitism" makes them resort to word juggling (e.g., that Arabs are Semites, too) which is tedious to analyze. Nevertheless, partly as the result of this juggling, the controversy over Shavarevich's writings is far from over. Recently, a volume of more than 500 pages has been brought out purporting to demonstrate that Shafarevich is not an anti-Semite.[9]

DOI: 10.1057/9781137338280

The combination of mathematics, politics, ultra-nationalism, and anti-Semitism was present in Moscow as early as the beginning of the twentieth century, and future historians may endeavor to trace how this tradition was passed down within Moscow mathematician circles. The case of Shafarevich cannot be adequately appraised without considering the political engagement of his distant predecessors, whether he is aware of them or not: the professional atmosphere that informed his views had a history, and this history started at the turn of the past century.[10]

The ideology of the Moscow mathematical circle of that period is thus not extinct. It continues to exist in one form or another, and this effects in retrospect our perception of the texts where we encounter it. Odd writings emanating from Moscow mathematicians of former times are not merely precious historical material sealed off in the past. They help to explain the peculiar combination of mathematics and anti-Semitism that still persists.

3

Due to the great achievements of Russian mathematicians in the twentieth century, the Moscow Mathematical Society is known all over the world. Numerous works have been written on the history of the Society. Very naturally, it has been a field for historians of mathematics. It is equally natural that their studies have touched on more than just the history of mathematics. Owing to the singular preoccupation of some Moscow mathematicians with the idea of making their discipline the foundation for all human knowledge, the materials related to the Society are often valuable for historians of a number of other subjects, such as philosophy, education, sociology, and theology. The journal *Istoriko-matematicheskie issledovaniia* [*Historical-mathematical studies*], regularly publishing documents and research on Moscow mathematicians, turns out to be an unexpectedly valuable source for students of Russian culture at the turn of the twentieth century. Much of what was printed there has helped me in the present study, even though my approach is entirely different, being focused on a theme which has never been given much thought or attention.

The main fact that the present book examines is very well-known. The staunch monarchism of Moscow mathematicians is frequently mentioned in studies, which may have contributed to the general neglect of

DOI: 10.1057/9781137338280

the subject, by its acquiring a deceptive air of familiarity, as if common knowledge. At a conference on the intellectual framework of modern mysticism, held in St. Petersburg in September 2010, Loren Graham, one of the authors of a recent book on Moscow mathematics, *Naming Infinity: A True Story of Religious Mysticism and Mathematical Creativity* (2009), expressed his surprise at my attempt to reconstruct the ideology of Bugaev's circle, saying that their monarchism was among the first things that he had heard about them. As usual, the commonplace word "monarchism" conceals its fluctuating historical meanings, and glosses over the different kinds of monarchism. One of the aims of my book is to reconstruct those aspects of the Moscow mathematical circle's monarchism that are of specific ideological significance. The picture obtained as a result will be very different from present conceptions about the milieu of Moscow mathematicians.

A trivial idea of their views is exemplified by the book of Graham and Kantor. The book aims at reconstructing an abstruse aspect of the Moscow mathematicians' intellectual divagations. We learn that certain important mathematical conceptions developed by the disciples of Bugaev, the brilliant mathematicians Nikolaĭ Nikolaevich Luzin (1883–1950) and Dmitriĭ Fedorovich Egorov (1869–1931) reflect their involvement with the mystical movement of onomatodoxy, or Name Worshiping [imîaslavie]. This involvement served as one of the pretexts of their persecution under the Soviet regime: they were accused of mixing mathematics with mysticism. Egorov died in exile. Pavel Aleksandrovich Florenskiĭ (1882–1937), another disciple of Bugaev, was shot. Luzin narrowly escaped the same fate. The charges pressed against them sound as fantastic as many others of the same period. But, as *Naming Infinity* shows us, in a way they were not unfounded. These mathematicians were indeed mystics, and their mysticism was linked with their mathematics.

The neglect by Graham and Kantor of one vital theme in their book is disappointing. Although the topic of mysticism forced the authors to reconstruct areas of thought far beyond mathematical problems, they omitted the relationship between the mysticism of Moscow mathematicians and their politics.[11] However, modern mysticism is rarely separable from political thinking, and therefore is difficult to comprehend without reference to it.[12] Moscow mathematicians were no exception. The description of the intellectual atmosphere of Bugaev's circle proposed in *Naming Infinity* is thus inadequate.

DOI: 10.1057/9781137338280

This omission of politics reflects a general trend in the representation of the Moscow Mathematical Society. Political issues are analyzed only in connection with its later history. A good example is afforded by an excellent book *Delo akademika Nikolaîa Nikolaevicha Luzina* ([*The case of the academician Nikolaï Nikolaevich Luzin*], 1999), which contains rich archival material concerning the ideological campaign of 1936 directed against Luzin and its political context. The material on the history of the Moscow Mathematical Society of the early twentieth century is of no less importance for the history of ideology. Yet the historical data is never considered from this point of view, as if politics became relevant to the affairs of the Mathematical Society only with the coming of the Soviet regime, which imposed its absurd ideology on scientists otherwise worlds away from political issues.

Mysticism was not the only charge leveled against the Moscow mathematicians in Soviet times. They were also accused of subscribing to the ideology of the Black Hundred.[13] The latter was the extreme right-wing monarchist organization involved in pogroms and other acts of violence. Just like the accusation of mysticism, probably even more so, this accusation appears at first glance to be absurd. In writings on the Moscow Mathematical Society we find almost nothing about its connection with any related ideology.[14] Thus nothing blurs the beautiful image recurrent in works on the Society: the image of a brilliant scientific community embodying the general flourishing of Russian culture at that time. As they consider the moral character of their heroes, Graham and Kantor mention Florenskiĭ's attitude towards the Jews in a very characteristic way. We are told that Florenskiĭ "was probably, in some of his writings, anti-Semitic," which implies that even if there are some anti-Semitic remarks to be found in his works, his general views were different.[15] *Naming Infinity* is a laborious attempt at white-washing a rather violent ideology as enlightened mysticism. One of the obvious factors behind such representations of the Moscow Mathematical Society is the memory of the macabre ideological campaigns of the late 1920s and 1930s. "The most extreme obscurantism [mrakobesie] under the guise of pure science" was prominent on the list of the characteristic features of the so-called Moscow philosophic-mathematical school in the Soviet ideological discourse. In the post-Soviet period the evaluations were reversed with even less factual evidence.[16]

It is not my purpose to deconstruct representations of Bugaev's circle, but the more one studies the history of the Moscow Mathematical

DOI: 10.1057/9781137338280

Society, the less one can preserve such an attractive picture of its activities outside its professional field. My purpose is to demonstrate that the ideology of some Moscow mathematicians is valuable for a historian of ideas less as part of a cultural renaissance, and more as a tangle of symptomatic delusions. By unraveling them we get into a strange intellectual universe where allusions to an imaginary Pythagorean tradition are combined with the most pressing issues of contemporaneous politics. The accusation that the Moscow mathematicians shared the Black Hundred's ideology was not groundless. Moreover, just as the mysticism of some mathematicians was deeply connected with their scientific ideas, there existed in the same circle a close link between extremely reactionary political views and mathematics.

A considerable part of this book is going to deal with anti-Semitism in the latter half of the nineteenth and beginning of the twentieth century. Although the role of anti-Semitic ideology in the culture of that time is generally recognized, much work remains to bring this general recognition to bear on individual facts that may elude our understanding if we fail to notice their deep immersion in this ideological climate.[17] Merely a general notion of this climate cannot give us the knowledge of the separate elements forming it. At the same time, neglecting these elements would reduce certain crucial beliefs to the status of commonplaces, or epiphenomena of no consequence.

While pointing out that anti-Semitism had vital significance for members of the Moscow philosophic-mathematical school, I was sometimes asked, "What was the specificity of their anti-Semitism?" The mystical fear of the Jews was widely abroad.[18] What was specific about Moscow mathematicians was their "mathematical" ideology, anti-Semitism being one of its key elements. Upon dismissing the latter element as not "specific," the full significance of the former as a universalistic ideology would be lost. Conversely, the writings of the Moscow mathematicians bring home the extent to which anti-Semitic beliefs were pervasive in contemporaneous culture, deeply pervading the most unlikely issues. Not only philosophical and political speculations could be defined by racial and religious prejudices, but also discussions of mathematics, of its history and social role.

Although most of what follows is devoted to Moscow mathematicians, I never lose sight of Belyĭ. Bugaev was a formative influence both on the Moscow Mathematical Society, and on Belyĭ himself (a prototype for Ableukhov the younger of the novel). Therefore, the parallels between

DOI: 10.1057/9781137338280

the ideological views of Bugaev's circle and those reflected in Belyï's work, most importantly in *Petersburg*, are illuminating. Moreover, since one of the key motifs of the novel is the parallel between Ableukhov the elder and Ableukhov the younger, my comparison of the ideology of Bugaev's circle with Belyï's views, should be relevant to the thematics of the novel.

The character of the "little senator" of *Petersburg* being a tour de force in many respects (including the amazing wealth of topical and historical allusions it contains), it also provides a convenient framework for a discussion of Bugaev's circle. My research on Moscow mathematicians prompted by the novel led to a fuller appreciation of Apollon Ableukhov's remarkably comprehensive character, his name itself being a complex topical allusion. Hence the relevant constituents of the senator's name and surname shall be scrutinized in further parts of this study.[19]

Since *Petersburg* is relatively little known outside Russia, one should add that it is undoubtedly one of the greatest novels of the past century.[20] Thus, even its smallest detail is entitled to close attention. The attempt to reconstruct a school of thought, using a literary motif not only as a point of departure, but as a constant frame of reference, is in this case not disproportionate. The more so for the geometrical passion of Ableukhov the elder is a valuable pointer to an ideology that perfectly illustrates the abuse of mathematics as described by the first historian of the discipline, Jean-Étienne Montucla (1725–1799) in terms applicable to our subject: "Les Pythagoriciens en montrerent autrefois l'exemple <...>; mais il est des modernes qui ont tellement enchéri sur ces visions creuses, qu'il n'est aucun Pythagorien qui ne leur eut cédé le pas <...>."[21]

The intricate symbols of *Petersburg* offer a synthesis of mathematics as a "state science" as conceived by the Moscow "school," and of racial mysticism as a key to understanding all social phenomena. To adequately interpret *Petersburg*, one has to trace its symbols back to their ideological components, including those related to the thought of the Moscow "school." Conversely, the novel may serve as a key to the nebulous political fantasies of the Moscow mathematicians.

Notes

1   During Belyï's lifetime, *Petersburg* came out in three different Russian versions: the first, completed in 1913, appeared in three installments in

DOI: 10.1057/9781137338280

the literary miscellany *Sirin* (1913–1914, reprinted as a separate edition in Petrograd in 1916); the second, considerably abridged, was brought out in Berlin by the Russian publishing house "Èpokha" in 1922; the third by "Nikitinskie subbotniki" in Moscow in 1928.

Besides, Belyĭ specially revised the first version for the German translation of *Petersburg*, published by Georg Müller press in Munich in 1919.

I used the "Sirin" text edited by L. K. Dolgopolov: Andreĭ Belyĭ, *Petersburg* (Sankt-Peterburg: Nauka, 2004). Since there are no considerable differences in what concerns the theme of geometry between the "Sirin" and the "Berlin" versions, I shall use the English translation of the latter by Robert A. Maguire and John E. Malmstad, with slight variations: Andrei Bely, *Petersburg* (Bloomington – London: Indiana University Press, 1978).

2   See Appendix 1.

3   V. Kagan, "Bugaev Nikolaĭ Vasil'evich," in *Bol'shaia Sovetskaia Èntsiklopediia*, T. 7 (Moskva: Gosudarstvennoe slovarno-èntsiklopedicheskoe izd. "Sovetskaia Èntsiklopediia", OGIZ RSFSR, 1927), 770.

4   The political views of Belyĭ's father and their relationship to those of Belyĭ, which is of vital import for the understanding of the latter's works, have never been studied. On Bugaev's mathematical conceptions as important for Belyĭ, see, in particular, works by Lena Szilard: *Germetizm i germenevtika* (Sankt-Peterburg: Izd. Ivana Limbakha, 2002), 283–295; "'Novaia matematika' i 'filosofiia matematiki' v *Istorii stanovleniia samosoznaiushcheĭ dushi*: Aspekty aritmologii i kombinatoriki," *Russian Literature* 70: 1/2 (2011), 137–157.

5   Sergeĭ Polovinkin, ed., *Sergeĭ Aleksandrovich Nilus (1862–1929)* (Moskva: Izd. Spaso-Preobrazhenskogo Valaamskogo monastyria, 1995).

6   See, primarily, "O studencheskom matematicheskom kruzhke pri moskovskom matematicheskom obshchestve v 1902–1903 gg.," *IMI* 30 (1986), 148–158; "Moskovskaia filosofsko-matematicheskaia shkola (Obzor)," *Referativnyĭ zhurnal. Obshchestvennye nauki v SSSR. Seriia 3. Filosofiia* 2 (1991), 43–67; "Psikho-aritmo-mekhanik (filosofskie cherty portreta P. A. Nekrasova)," *Voprosy istorii estestvoznaniia i tekhniki* 2 (1994), 109–113. See also his book *P. A. Florenskiĭ: Logos protiv Khaosa* (Moskva: Znanie, 1989).

7   See Françoit Furet, *Penser la Révolution française* (Gallimard, 1978), 257–316.

8   See Chapter 3, Section "Dead soul of the world."

9   Krista Berglund, *The Vexing Case of Igor Shafarevich, a Russian Political Thinker* (Basel: Springer, 2012).

10   Along with Shafarevich, who has written a second anti-Semitic volume entitled *Trekhtysiacheletniaia zagadka* (Sankt-Peterburg: Bibliopolis, 2002), the outstanding Moscow mathematician Lev Semenovich Pontriagin (1908–1988) is an obvious example of adherence to the same school of thought. See his *Zhizneopisanie L'va Semenovicha Pontriagina, matematika, sostavlennoe im samim* (Moskva: IChP "Prima B," 1998).

DOI: 10.1057/9781137338280

11 There is an account of the history of the persecution of the Name Worshipers both before and after the October revolution, but still there is almost nothing about the political component of their mystical beliefs.

12 A striking example of this link is Mme Blavatsky's letter to a general of the Russian Imperial Corps of the Gendarmes, in which she proposed her services as a spy, indicating that her involvement with spiritualism placed her in an advantageous position for obtaining secret information (B. L. Bessonov and V. I. Mil'don, eds, "Pis'mo Blavatskoĭ," *Literaturnoe obozrenie* 6 (1988), 110–112; this letter was brought to my attention by Omry Ronen).

13 See, for example, S. S. Demidov and B. V. Levshin, eds, *Delo akademika Nikolaĭa Nikolaevicha Luzina* (Sankt-Peterburg: RKHGI, 1999), 18, 21, 35, 65, 82, 96, 99 etc.

14 Thus, S. S. Demidov mentions the Soviet characterization of the Moscow mathematicians, which indicated their connection with the Black Hundred, as a "label [ĭarlyk]": Sergeĭ S. Demidov, "O matematike v tvorchestve P. A. Florenskogo," in Michael Hagemeister and Nina Kauchtschischwili, eds, *P. A. Florenskiĭ i kul'tura ego vremeni. P. A. Florenskij e la cultura della sua epoca. Atti del Convegno Internazionale Università degli Studi di Bergamo 10–14 gennaio 1988* (Marburg: Blaue Hörner Verlag, 1995), 177.

   The important exception is Oscar Sheynin's studies on P.A. Nekrasov. See Oscar B. Sheynin, "Nekrasov's Work on Probability: The Background," *Archive for History of Exact Sciences* 57 (2003), 343; O.B. Sheĭnin, "Publikatsii A.A. Markova v gazete 'Den' za 1914–1915 gg.," *IMI* 34 (1993), 196; M.V. Chirikov and O.B. Sheĭnin, "Perepiska P.A. Nekrasova i K.A. Andreeva," *IMI* 35 (1994), 124. For the English translation of these articles, see http://www. sheynin.de/download/2_Russian%20Papers%20History.pdf, 63–69, 70–82. For Nekrasov's reactionary views, see also materials collected, translated and commented by Oscar Sheynin in P.A. Nekrasov, *Theory of Probability*: http:// www.sheynin.de/download/5_Nekrasov.pdf, 2, 55–64.

15 Loren Graham and Jean-Michel Kantor, *Naming Infinity: A True Story of Religious Mysticism and Mathematical Creativity* (Cambridge, MA – London: The Belknap Press of Harvard University Press, 2009), 196. In the French edition of this book the phrase was changed, and the passages on anti-Semitism as a marked feature of the Bugaev circle were added: *Au nom de l'infini: une histoire vrai de mysticisme religieux et de création mathématique* (Paris: Belin, 2010), 252, 90–91, 95–96. These changes followed our correspondence with Jean-Michel Kantor, not mentioned in the text. On Florenskiĭ's anti-Semitism, see Michael Hagemeister's works cited in n. 18, and in Chapter 4, n. 46.

16 As an example one may cite the following presentation of P.A. Nekrasov's views: "On account of Nekrasov's close connection with the Tsarist régime <...>, his deeply conservative and religious views, his somewhat erratic

mathematical writings from about 1900, his espousal of a connection between religion and mathematics, and of course his conflicts and controversies with Markov, the better probabilist, led the Communist régime to denigrate him as an eccentric and brand him as a reactionary" (Eugene Seneta, "Statistical Regularity and Free Will: L.A.J. Quetelet and P.A. Nekrasov," *International Statistical Review* 71: 2 (2003), 320). The point is, however, that Nekrasov was indeed a reactionary, and the one of the deepest dye.

17   The significance of this topic for our understanding of contemporary Russian culture has been amply demonstrated in the works by Mikhail Bezrodnyĭ, "O 'iudoboĭazni' Andreîa Belogo," *Novoe literaturnoe obozrenie* 28 (1997), 100–125; Omry Ronen, "'Èto'" in his *Shram. Vtoraîa kniga iz goroda Ann* (Sankt-Peterburg: zhurnal "Zvezda," 2007), 63–73, and "Poedinki" in his *Chuzheliubie. Tret'îa kniga iz goroda Ann* (Sankt-Peterburg: zhurnal "Zvezda," 2010), 199–215; and Mikhail Zolotonosov, *"Master i Margarita" kak putevoditel' po subkul'ture russkogo antisemitizma (SRA)* (S.-Peterburg: INAPRESS, 1995).

18   The following words of a contemporary reviewer of *Petersburg* aptly characterize this frame of mind: "<...> through the frail texture of the visible reality and of the normal daily consciousness a different reality shows <...> in which dark roots of history come to light, and there in that different reality Semitic features come through <...> which in morbid delirium, in hallucination are taking shape <...> of the fatal signs of death" (L.S. Vygodskiĭ, "Literaturnye zametki ('Peterburg,' roman Adreîa Belogo. 1916 g.)," *Novyĭ put'*, December 11, 47 (1916), st. 31–32). The same foundations of history "show through" in Florenskiĭ's preface to the rampantly anti-Semitic collection of articles *Izrail' v proshlom, nastoîashchem i budushchem [Israel in the past, present and future]*: "There is hardly anybody who does not see that the Jewish question is the world question, and, moreover, is the central question of world history. It is in this knot that countless and tangled threads of history come together." See "Predislovie," in *Izrail' v proshlom, nastoîashchem i budushchem* (Sergiev Posad: Izd. "Religiozno-filosofskoĭ Biblioteki," 1915), 5. The preface is anonymous. On the authorship of Florenskiĭ, see Michael Hagemeister, "Pavel Florenskij und der Ritualmordvorwurf," in Michael Hagemeister and Torsten Metelka, eds, *Appendix 2. Materialien zu Pavel Florenskij* (Berlin u. Zepernick: Kontext, 2001), 69.

19   Apollon, it should be noted, is an accepted first name among the Russian nobility, and three well-known nineteenth-century poets bear it (in the descending order of merit: Apollon Grigor'ev, Apollon Maĭkov, and Apollon Korinfskiĭ). It is, however, quite rare among Russian statesmen and high officials. The two who come to mind are privy councilor Apollon Davydovich Lokhvitskiĭ, the Eniseĭ governor in the 1880s, and Apollon Konstantinovich

DOI: 10.1057/9781137338280

Krivoshein, the last minister of transportation under Alexander III, eventually accused of graft and forced into retirement.

20 One may invoke the authority of Vladimir Nabokov, who included *Petersburg* in his list of the greatest literary masterpieces of the past century, after Joyce's *Ulysses* and Kafka's *Metamorphosis* (closing the list and placed immediately after *Petersburg* finds itself "the first half of Proust's fairy tale *In Search of Lost Time*"). See Vladimir Nabokov, *Strong Opinions* (New York: Vintage International-Vintage Books—A division of Random House, Inc., 1990), 57; cf. Robert A. Maguire and John E. Malmstad, "Translators' Introduction," in Bely's *Petersburg*, VIII.

21 Montucla *Histoire des mathématiques*, T. 1 (Paris: Ch. Ant. Jombert, 1758), 37 (h).

DOI: 10.1057/9781137338280

1
Origins of the Moscow "School": N. V. Bugaev

Abstract: *Materials preserved in N.V. Bugaev's archives allow us to explore the origins of the ideology of the Moscow philosophic-mathematical school and, in particular, the anti-Semitic elements of this ideology.* Bugaev's Foundations of Evolutionary Monadology (1893), *as well as* Belyĭ's Symbolism *(1910), are considered within the framework of the European "Aryan renaissance."*

Svetlikova, Ilona. *The Moscow Pythagoreans: Mathematics, Mysticism, and Anti-Semitism in Russian Symbolism.* New York: Palgrave Macmillan, 2013. DOI: 10.1057/9781137338280.

DOI: 10.1057/9781137338280

Bugaev's library

"Pythagore le pensait déjà: les mathématiques sont á la base de tout." This phrase, which I recently came across in a Russian schoolbook for French, sounds so banal that it could pass without remark. This makes it difficult for us to pick up such a formula appearing in a different historical context where it in fact constituted a matter of intense reflection. This marks a sharp difference: for us such a phrase is almost empty of meaning, being a part of common discourse, whereas at the time of its historical reemergence it acquired new significance now eluding us. An apparent coincidence in form (the Moscow mathematicians would have subscribed to the phrase from the schoolbook in its entirety) masks a profound divergence: a truism in one case, and a matter of deep reflection in the other. In the case of Bugaev's disciples, the phrase's revival was partly contingent on the political element of the Pythagorean tradition, both historically (the Pythagorean school actively participated in political life) and logically: if we are talking of mathematics as a foundation of all human knowledge, that implies politics as well.

The papers from Bugaev's archives point to the origins of the renewed interest in the idea of a universal significance of mathematics.[1]

1

One of the most valuable documents preserved in Bugaev's archives is the catalog of his library.[2] It was drawn up at the beginning of the 1880s, when the collection ran to 1600 volumes, and Bugaev still had 20 years ahead of him to add to the collection. There is, however, every reason to believe that the library's structure remained unchanged. The catalog consists of 28 sections, ranging from mathematics and other sciences to sociology and psychology.

The last section of the catalog is that of the belles-lettres. Not only is it the very last, but also the smallest section, comprising only seven books—a strange assortment at that, with Dante and Goethe juxtaposing authors unknown even to most historians of literature.[3] This seems to point to Bugaev's lack of any pronounced interest in literature, and also has implications for Belyĭ, who grew up in this library. He manifested the same passion as his father for a broad spectrum of knowledge, according to the lists of books he read, among which works of fiction and poetry

DOI: 10.1057/9781137338280

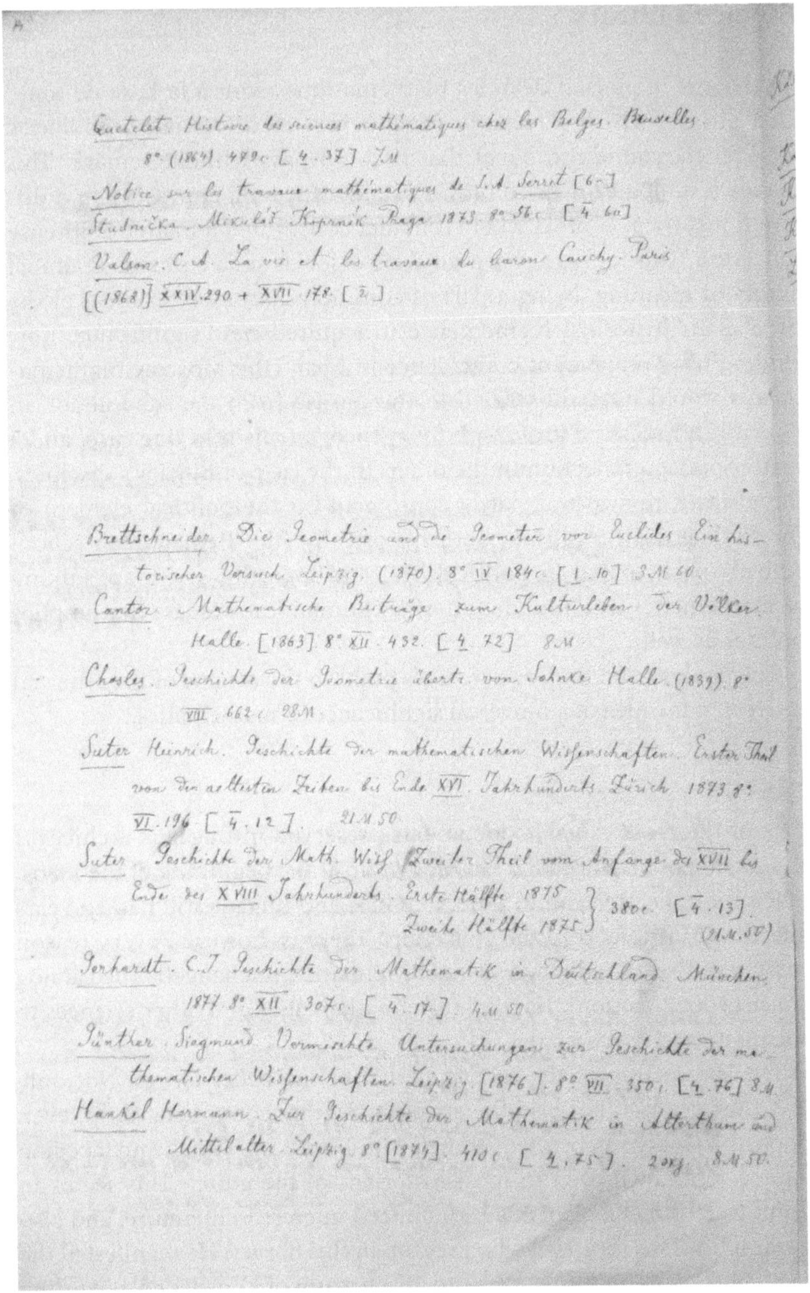

Quetelet. Histoire des sciences mathématiques chez les Belges. Bruxelles
8° (1864). 479 c. [4 . 37] 7 M
Notice sur les travaux mathématiques de J.A. Serret [6 –]
Studnička. Mikuláš Koprník. Praga 1873. 8° 56 c. [4 60]
Valson. C.A. La vie et les travaux du baron Cauchy. Paris
[(1868)] XXIV.290 + XVII 178. [2]

Brettschneider. Die Geometrie und die Geometer vor Euclides. Ein historischer Versuch. Leipzig. (1870). 8° IX 184 c. [1 . 10] 3 M 60
Cantor. Mathematische Beiträge zum Kulturleben der Völker. Halle. [1863] 8° XII. 432. [4 72] 8 M
Chasles. Geschichte der Geometrie übertr. von Sohnke. Halle. (1839). 8° VIII. 662 c. 28 M
Suter Heinrich. Geschichte der mathematischen Wissenschaften. Erster Theil von den ältesten Zeiten bis Ende XVI. Jahrhunderts. Zürich 1873 8° VI. 196 [4 . 12]. 21 M 50
Suter. Geschichte der Math. Wiss. Zweiter Theil vom Anfange des XVII bis Ende des XVIII Jahrhunderts. Erste Hälfte 1875, Zweite Hälfte 1875 } 380 c. [4 . 13]. (21 M. 50)
Gerhardt. C.J. Geschichte der Mathematik in Deutschland. München. 1877. 8° XII 307 c. [4 . 17]. 4 M 50
Günther Siegmund. Vermischte Untersuchungen zur Geschichte der mathematischen Wissenschaften. Leipzig [1876]. 8° VII 350 c. [4 . 76] 8 M
Hankel Hermann. Zur Geschichte der Mathematik in Alterthum und Mittelalter. Leipzig 8° [1874]. 410 c. [4 . 75]. 2 экз. 8 M 50.

FIGURE 1.1 *A page of the catalog of N. V. Bugaev's library (ORK i R NB MGU)*

DOI: 10.1057/9781137338280

are comparatively rare.[4] Thus, his father's library catalog indicates why commentary to Belyï's work involves us in intellectual history going far beyond the realm of literature.

2

Despite the apparent incongruity of some sections, there is nothing arbitrary about the catalog taken as a whole. Bugaev seems to have collected his library in a systematic way. It is not the library of a mathematician who is focused exclusively on his field, nor does it seem to be one of a dilettante jumping at random from one interest to another.

Among Bugaev's papers, there is a copybook containing his notes taken from the second edition of Montucla's classical *Histoire des mathématiques* (1799–1802).[5] It is important that Bugaev includes in his notes what many mathematicians would have omitted as extraneous to mathematics as such. In particular, he is attracted to what Montucla writes in the beginning of his history about the great respect that mathematics enjoyed amongst the ancients. There are notes (admittedly not very accurate) revealing Bugaev's interest in the following passage:

> Les mathématiques furent toujours accueillies avec une estime singulière par les philosophes les plus respectable de l'antiquité. Nous remarquerons en effet que tout ceux dont la doctrine et les mœurs furent les plus parfaites, cultivèrent ces connoissances, ou du moins en firent cas. <…> Ainsi pensèrent, pour ne citer que les plus célèbres, *Thalès*, *Pythagore*, *Démocrite*, *Anaxagore*, et tous les philosophes des écoles *Ionienne* et *Italique*; *Platon* enfin, *Xenocrate*, *Aristote*, &c. Personne n'ignore que les premiers de ces philosophes contribuèrent de tous leurs soins aux progrès qu'elles firent dans la *Grèce*; que *Platon* fut un des plus habiles Géomètres de son temps, et que ses ouvrages sont remplis de traits honorables pour les mathématiques. Quel cas ne témoignoit-t-il pas faire de la géométrie, lorsque questionné sur les occupations de la divinité, il répondit qu'elle *géométrise continuellement*, c'est-à-dire, sans doute, qu'elle gouverne l'univers par des lois géométriques?[6]

This passage is abbreviated in Bugaev's notes, along with Montucla's reference to mathematics as the fundament of the education and philosophy of the ancients, and Hippocrates instructing his son about the usefulness of mathematics for medicine. Bugaev also copied out a list of later philosophers who were also mathematicians, as cited by Montucla.[7] Clearly, these are the notes of an enquirer interested not only in mathematics

DOI: 10.1057/9781137338280

but also in its interrelations with other disciplines (the universality of Leibniz fascinated Bugaev for years[8]), and who is attracted by the idea of the primary importance of mathematics among all the disciplines.

Bearing this in mind, Bugaev's library was an accurate expression of his intellectual preferences. It is unlikely that in collecting books he merely strove after general erudition, without being guided by his thinking about mathematics. It would rather seem that the variety of subjects to be found in the catalog of his library reflected his thoughts on mathematics: ruminations on the universal significance of mathematics presupposed interest towards other disciplines. From the point of view of a "universal mathematician," all other disciplines were linked to his own domain, meaning that all of them were to become mathematical:

> To find measure in the domain of thought, will and feeling—this is the task of the contemporary philosopher, politician and artist.[9] <...> From the sphere of undefined, limitless, animal instincts, man is striving, with the help of number and measure, to rise to the ideal condition of having full power over outer and inner nature, of establishing harmony and esthetical feeling in every manifestation of the human spirit.[10]

Among Bugaev's teachers, I have found only one manifestation of comparable beliefs in the power of mathematics. In his speech *O vliĭanii matematicheskikh nauk na razvitie umstvennykh sposobnosteĭ* [*On the influence of mathematical sciences on the development of mental faculties*], delivered in 1841, Nikolaĭ Dmitrievich Brashman (1796–1866) referred to some of the above cited examples of the importance of mathematics, adduced by Montucla and copied by Bugaev.[11] There may have been some discussions in the Moscow mathematical circle that influenced Bugaev's interest in mathematics as a fundamental discipline of universal import. However, apart from Brashman's short speech there is little evidence as to their nature.[12]

3

In the second half of the nineteenth century, the idea of the potential universality of mathematical knowledge does not appear to have excited much general or professional interest. If, for comparison, we open the article *Géometre* by d'Alembert in the *Encyclopédie*, published a year before the above-mentioned book by Montucla, we will find a similar train of thought enthusing over the universal relevance of mathematics. The trend established in the seventeenth century of charging mathematics with meanings going beyond mathematics itself, to see it as a foundation for education and

DOI: 10.1057/9781137338280

a vehicle for moral and social perfection, that is, to impart an ideological role to mathematics, was still well alive in the eighteenth century. According to d'Alembert, the study of geometry precedes the spread of enlightenment: "C'est peut être le seul moyen de faire secoüer peu-à-peu à certaines contrées de l'Europe, le joug de l'oppression et de l'ignorance profonde sous laquelle elles gémissent."[13] Thus, mathematics acquires an ideological character, perceived not merely as a particular scholarly field, but as closely linked with ideas about moral and political improvement of society.

In the nineteenth century, interest in mathematics as a universal discipline, with implications for political ideology, was rare among professional mathematicians.[14] Celebrations of mathematics as "the only science leading to all possible knowledge" mainly occurred in treatises on mystical mathematics.[15]

It was perhaps not any contemporary intellectual trend that defined Bugaev's tastes, but rather his intense reading on the history of mathematics and philosophy. A considerable section of his library comprised books on the history of mathematics,[16] and it was here that the idea of mathematics as the groundwork for the entire body of learning was expounded, particularly in connection with the views of the ancient Greeks. The very first pages of such volumes featured celebrations of mathematics as the "core of all knowledge [den Kern aller Kenntnisse]" and "the bearing pillar or the main support of all the world wisdom [die Grundpfeiler oder Hauptstützen der gesammten Weltweisheit]."[17] It was only natural to allude to such characterizations when discussing Pythagoras, who was not only "the father of mathematics," but also the creator of the first all-embracing philosophical system placing mathematics at the base of all disciplines, including philosophy.[18]

There is thus reason to believe that the ideology of the Moscow Mathematical Society—reviving the idea of mathematics as a universal science and thus conjoining mathematics with politics—owed much to Bugaev's erudition. This conclusion is further substantiated by the work of Bugaev's disciple Pavel Alekseevich Nekrasov (see below), who reduced this line of thought to an absurdity more compatible with speculative knowledge than a living intellectual tradition.

Positivism

Two often overlooked factors concerning universality rather than mathematics further contributed to the ideas of Bugaev's circle.

DOI: 10.1057/9781137338280

The first such factor was positivism, the real historical meaning of which is often unjustifiably reduced to the notorious question of "facts." As a result, while the word "positivism" is constantly adduced in studies on this period (including those on the Moscow Mathematical Society), this reflects only its frequent occurrence in works written at the time. Its real historical implications are rarely examined.[19] One of them was positivistic universalism. The voluminous tracts of Comte or Spencer dealing with all imaginable subjects exemplified the idea of the all-encompassing scientific worldview, and this quality in itself was probably as influential as, and counteracted, the explicit statements contained therein as to the necessity of scientific specialization. Mathematicians envisaging a new universal outlook were bound to have positivistic works in mind, not only because the latter were very popular, but also because they were the very embodiment of universality.

Moreover, the discipline placed by Comte in a similar position to that of mathematics, according to the views of Bugaev's circle, was sociology, which was also of much importance to Moscow mathematicians. Bugaev did not write on sociology, but there were a conspicuous number of books on the subject in his library.[20] This interest is likely to have been associated with Bugaev's mathematical studies, as works by his disciple show. For Nekrasov, mathematics came to be closely connected with sociology owing to his research on theory of probability and statistics. This meant that, for this circle, positivism was a still more inevitable rival. Positivist effort was not totally different: it was aimed at constructing a universal system of knowledge in which sociology was given prominence. One of the fatal flaws of positivism, from the point of view of the Moscow mathematicians, was its insufficient use of mathematics. In their view, positivism's neglect of mathematics led to the creation of a system misrepresenting both the whole of knowledge (in particular, the erroneous stress on facts could be avoided by considering the role of theoretical presupposition in geometry), as well as its most important subject, sociology, which the Moscow mathematicians believed had to be based on mathematics.[21]

This partly accounts for numerous allusions to positivism in the works of the Moscow mathematicians. It was a much more organic part of their intellectual world than one may assume from their attacks on positivistic "facts" and suggested "materialism."

In fact, even the mystical and religious tendencies of this circle are not simply a refutation of positivism. Such tendencies would rather appear to have owed something to positivists, or authors regarded as such at the time. It was enough to open the then well-known *History of the Inductive*

DOI: 10.1057/9781137338280

Sciences (1837) by William Whewell to find ample information on mysticism, theosophy, astrology, magic, and the like.[22] Readers of such books—and the history of science was one of the favorite subjects for positivists—were better versed in these matters than one expects today.[23] Attitudes to them may have sometimes been more sympathetic, the more skeptical readers' sources were. Considering the section of the catalog of Bugaev's library entitled "Cosmology. Theology,"[24] it is reasonable to suppose that Bugaev also shared this common positivistic interest in the history of scientific knowledge.

Navigating the anti-positivistic polemics regularly adduced in studies on Bugaev's circle, one is confronted with the recurrent problem of historical research, formulated variously on numerous occasions: the contrast between the clear-cut tendencies as presented in historical studies (often for the very good reason of better classification) and the confused cross-currents of actual history. We will encounter in our subject combinations most incongruous from the point of view of accepted classifications, whereby elements generally supposed to be utterly incompatible were in fact inseparable, if only due to a mutual repulsion that defined their respective meanings.

Psychology

The second factor contributing to the universalist outlook of Bugaev's circle was psychology. At the turn of the twentieth century, it was still widely believed that psychology was the true fundament for all the disciplines.[25] Hence such works as Wilhelm Wundt's 10-volume *Völkerpsychologie* (1900–1920) encompassed a strikingly wide range of subjects. These works provided yet another example of a scientifically based universal knowledge and were an indispensable part of the mental equipment of the scholars at that time.[26]

Accordingly, there is a vast section on psychology in the catalog of Bugaev's library.[27] His copious notes on the subject have also been preserved.[28] Bugaev was among the first members of the Moscow Psychological Society, which became one of the most important intellectual centers in Moscow and published the first Russian philosophical journal, *Voprosy filosofii i psikhologii* [*Questions of Philosophy and Psychology*].

But there was a tighter link between the theoretical pursuits of Moscow mathematicians and psychology than the mere reference to their interest

DOI: 10.1057/9781137338280

in the subject, or to its widespread popularity, implies. At the beginning of the twentieth century the psychology of Johann Friedrich Herbart (1776–1841) was all but forgotten. The terms he had coined or made popular had come to be an important part of intellectual language, widely used not only in psychology, but in various fields of humanities that used psychology as their basis. Hence the frequent recurrence of such expressions as "light (or dark) field of consciousness," "threshold of consciousness," "suppression" etc. in works at that time, which now tend to be mistakenly interpreted as traces of Freudian influence, while in fact Freud was only one of those who used this language.[29]

Moscow mathematicians must have been particularly interested in Herbart, who had proposed the project of mathematical psychology. Herbartian psychology could therefore serve as a proof that even psychology, widely regarded as a fundamental discipline, was ultimately dependent on mathematics. Bugaev's aforementioned striving to "find measure in the domain of thought, will and feeling," taken up by his disciples, is to be regarded within this context.[30]

In connection with the leanings of Moscow mathematicians towards mysticism, one should also bear in mind that contemporary psychology was another vital factor behind the interest in mystical and occult traditions. Religious beliefs were regarded as important psychological phenomena to be analyzed. Bugaev was most probably attracted by this idea. In his notes there is a reference to the book by Louis Figuier, *Histoire du merveilleux dans les temps modernes* (1860).[31] The likely source of the reference was the review of this book published in *Otechestvennye zapiski* which strongly emphasized the importance of data concerning popular superstitions and religious beliefs to psychology.[32] The immediately subsequent reference in Bugaev's notes is to Éliphas Lévi's *Histoire de la magie: avec une exposition claire et précise de ces procédés, de ces rites et de ces mystères* (1860). It is not clear whether Bugaev read it, and if he read it what he thought of it, but the way the title emerges in his notes tells us something about how his curiosity about, and attraction towards, the occult tradition may have started.[33]

As is well known, the context of contemporary psychological research—an important framework motivating studies on religious beliefs—could also foster views not normally associated with a modern scientific outlook. Belyĭ's memoirs evoke "stories about ghosts" discussed at the flat of his father's friend and president of the Moscow Psychological Society Nikolaĭ Ĭakovlevich Grot (1852–1899).[34] The next president of

DOI: 10.1057/9781137338280

the Society and Bugaev's friend Lev Mikhaĭlovich Lopatin (1855–1920) was passionately interested in spiritualism.[35] Although in his memoirs Belyĭ depicts his father as being very skeptical on the matter, the latter's lectures testify to a different attitude:

> The first thoughts about antipodes, about the movement of the earth did not have any analogy with the then dominating notions and roused strong objections and arguments. Likewise, in our time the table rapping, slate writing etc., did not get common credit, at least among naturalists, in spite of various and numerous observations because they did not have any analogy with phenomena already known to us.[36]

Anti-Semitism

Anti-Semitism may seem totally foreign to the project of creating a universal science or worldview, and still more so to any universality based on mathematics. As for Bugaev himself, we do not have direct evidence that he linked his thinking about mathematics to his anti-Semitism. Yet materials conserved in his archives do shed light on this ideological aspect of the Moscow philosophic-mathematical school.

1

In his memoirs, one of Belyĭ's closest friends writes that Jews ["zhidy"] were Bugaev's "idée fixe" and that he always had with him a notebook where he put down "all the outrageous facts from the life of Israel."[37] This notebook has been lost, but there is a second notebook to be examined in this connection.

In it Bugaev recorded facts and events without any apparent connection between them, marking each entry with a number, so that only the numbers bring some order into the notes varying from such as "in China they do not know the use of milk" or "Marquise de Brinvilliers was a famous poisoner" to "the brothers Grimm founded the historical grammar."[38] On two instances the Jews are evoked. The number 3 runs as: "Meyerbeer and Heine are Jews," the number 15 as: "Mendelssohn, Rachel, Spinoza are yids."[39] The context is such that we cannot get much out of these notes, except that the Jews were of interest to Bugaev since quite early on (the notebook dates from the beginning of 1860s) and that this interest was not friendly.

DOI: 10.1057/9781137338280

2

Next, there is a letter by P. A. Nekrasov (dated May 11, 1894) from which we learn that the Moscow Mathematical Society had thanked the Grand Duke Sergeï for his attentiveness to its needs.[40]

The Grand Duke Sergeï Aleksandrovich (1857–1905) was one of the most prominent Russian nationalists. In 1891 he was appointed governor-general of Moscow. His functions included providing assistance to scientific and cultural organizations. In this particular case, though, his benevolence might have been reinforced by Bugaev's adherence to similar political views. Bugaev was a nationalist.[41] One of the remarkable consequences of his nationalism is that in the twentieth century knowledge of Russian was widespread among mathematicians all over the world. It was Bugaev who insisted on publishing the journal of Moscow Mathematical Society in Russian.[42]

There was also another point of similarity between the political positions of the Grand Duke and Bugaev. Once appointed governor-general in 1891 (the same year Bugaev was elected president of the Moscow Mathematical Society), the Grand Duke organized the mass expulsion of Jews from Moscow. His nationalism and patriotism were closely linked with anti-Semitism, which had become an almost inseparable part of the ideology of the Right, particularly in the wake of the assassination of Alexander II in 1881 that was widely attributed to Jews. Jews came to be regarded by conservative and especially ultra-conservative circles as the main enemies of the monarchy and the state.

It is within this frame of reference that one has to consider Bugaev's anti-Semitism. In his archives there is a list of students involved in the disturbances at Moscow University in 1901. It was drawn up at the request of Nekrasov, who was at the time the rector of Moscow University. Near some of the names Bugaev notes: "an Israelite [iudeï]," which is evidently an indication of the main instigators.[43] This is a manifestation of the same ideology that ten years later would become one of the central elements in his son's great novel: social disorder and revolutions are brought about by the Jews.

3

Looking at the invitation for an evening party given by the Grand Duke Sergeï, preserved in Bugaev's archive,[44] there is reason to believe it is not merely the insignificant trace of a social ritual. They shared certain views that may have provided a common ground for their conversations.[45]

DOI: 10.1057/9781137338280

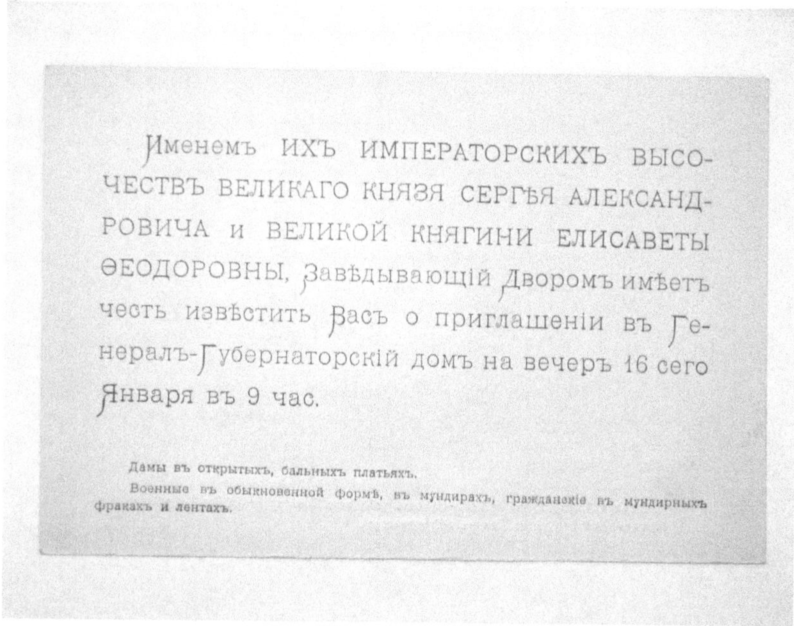

Именемъ ИХЪ ИМПЕРАТОРСКИХЪ ВЫСОЧЕСТВЪ ВЕЛИКАГО КНЯЗЯ СЕРГѢЯ АЛЕКСАНДРОВИЧА и ВЕЛИКОЙ КНЯГИНИ ЕЛИСАВЕТЫ ѲЕОДОРОВНЫ, Завѣдывающій Дворомъ имѣетъ честь извѣстить Васъ о приглашеніи въ Генералъ-Губернаторскій домъ на вечеръ 16 сего Января въ 9 час.

Дамы въ открытыхъ, бальныхъ платьяхъ.

Военные въ обыкновенной формѣ, въ мундирахъ, гражданскіе въ мундирныхъ фракахъ и лентахъ.

FIGURE 1.2 *The invitation for an evening party given by the Grand Duke Sergeĭ Aleksandrovich (ORK i R NB MGU)*

Following the Grand Duke's assassination in 1905, Bugaev's disciple P. A. Nekrasov dedicated a book to his memory. This book was partly read at the meeting of the so-called Union of the Russian people [Soîuz russkikh lîudeĭ)], an organization close to, and overlapping with, the Black Hundred movement.[46] The traces of the Grand Duke in Bugaev's archive, the first of them dating back to the beginning of Bugaev's presidency of the Moscow Mathematical Society, anticipate the Society's later history. They foreshadow the amazingly—for we are talking about mathematics—close interrelations between the Moscow Mathematical Society, under Bugaev's influence, and ultra-conservatism.

4

There is another interesting document in Bugaev's archives concerning our subject: a letter of thanks from Vladimir Andreevich Gringmut (1851–1907).[47] We learn from it that Bugaev had sent his philosophical tract *Osnovy èvolîutsionnoĭ monadologii* ([*Foundations of Evolutionary*

DOI: 10.1057/9781137338280

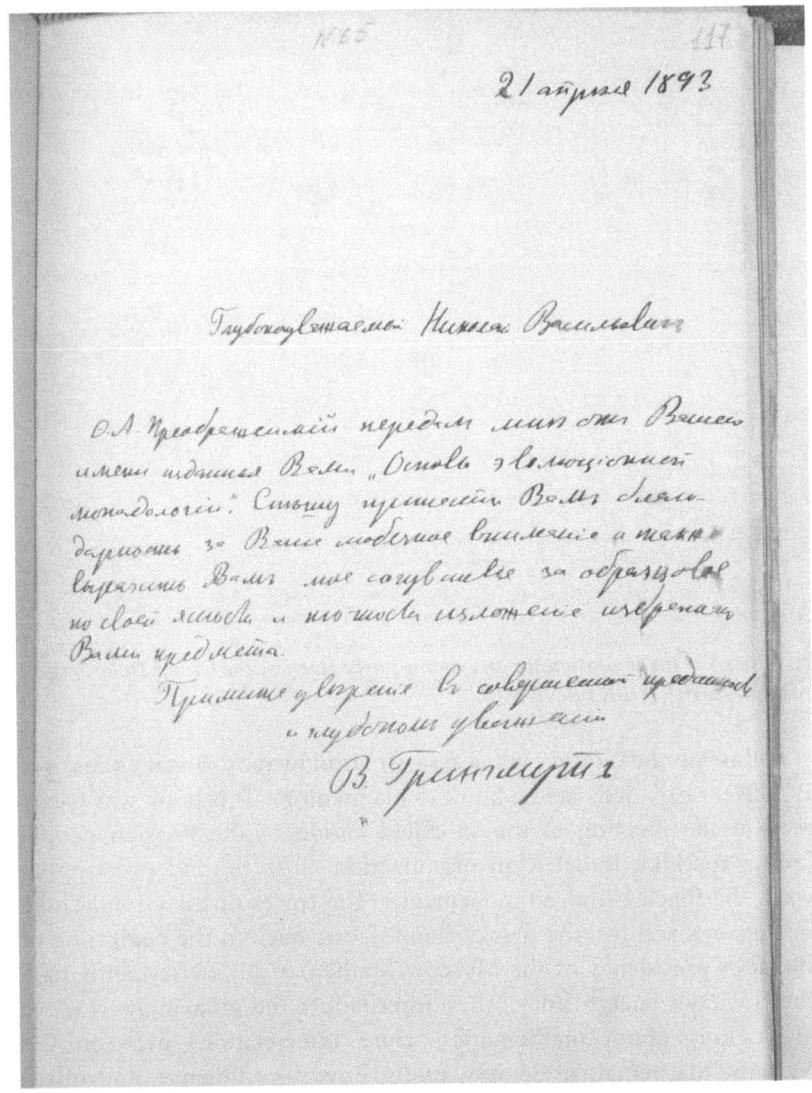

FIGURE 1.3　*A letter of thanks from V. A. Gringmut (ORK i R NB MGU)*

Monadology], 1893) to one of the future founders of the Black Hundred movement, whose *Rukovodstvo chernostotentsa-monarchista* ([*Guide of the Black Hundred Monarchist*], 1906) is still popular with contemporary Russian anti-Semites.[48]

DOI: 10.1057/9781137338280

Bugaev's little tract, intended as a development of Leibniz' *Monadologie*, and apparently of only antiquarian interest, also catches our attention in a somewhat different way: Bugaev's abstract and unexciting reasoning on the ordered and regulated life of monads submitting to perfect laws of measure and harmony, which nonetheless displays considerable enthusiasm for its subject, might have also contained a contemporary political message. Apart from its curious source of inspiration, of which we shall treat below, this booklet probably reflected, in its own way, an optimistic outlook characteristic of conservatives under the rule of Alexander III. This might have prompted Bugaev to send this barely readable tract to Gringmut, who was not a philosopher but (at that time) a well-known conservative journalist, soon to become the chief editor of the main conservative Russian paper *Moskovskie Vedomosti* [*Moscow Gazette*]. Such a connection between Bugaev's monadology and his conservative attitudes may seem sketchy owing to the absence of any evidence of intermediary links. But it becomes more plausible when we look at what the concept of "harmony," one of the key terms of this treatise, would develop into later, in the hands of some of Bugaev's pupils, whose obsession with "harmony" and "chaos" made fully fledged political categories out of the words.[49]

5

In connection with Bugaev's nationalism and conservatism one should also mention his pan-Slavic attitudes. He was a passionate reader of the journal *Slavîanskiĭ vek* [*The Slavic Century*].[50] Published in Austria, it had a strongly anti-Austrian platform and propagated pan-Slavic ideas. In Bugaev's archives we find an invitation to join the Slavic Supporting Society in Moscow [Slavîanskoe vspomogatel'noe obshchestvo v Moskve] which helped find funds for publishing this journal.[51] Bearing this in mind and examining the section on the history of mathematics in the catalog of Bugaev's library, one wonders whether his interest in Copernicus, as manifested by the presence of four books on the Pole (more than on any other scientist in the same section), was connected with his pan-Slavic sentiments.[52] As the works of his follower Baron M. F. Taube suggest, to be discussed in Chapter 4, in this context, even the passion for Leibniz, who was of Sorbian origins, might well have had elements of the same ideology.

DOI: 10.1057/9781137338280

A cult of India

1

In his aforementioned memoirs, Sergeĭ Solov'ev wrote: "<...> still more than by the Russian mode of life [russkim bytom], Bugaevs' apartment was saturated with the spirit of India. All the family was absorbed in reading Blavatsky, Borîa [Bugaev] initiated me in the mysteries of yogi and of spiritism <...>."[53] From Belyĭ's memoirs we know that his father had written a libretto for an opera *Buddha* which was shown to and appreciated by Richard Wagner.[54] In the context of what has already been said, these remarks seem to testify to a coherent system of views reflecting the general European trend whereby unworthy Semites were contraposed to the noble Aryans. Bugaev's anti-Semitism, closely connected to the politics of that time, was thus apparently supported by widespread historical fantasies. What is interesting is that such fantasies must have influenced his ideas about the history of mathematics.

Bugaev's library contains clues as to what he knew about the history of mathematics, especially if we also consider works probably acquired after the catalog itself was completed. The first book in the section "The history of mathematics" is *Istoricheskiĭ ocherk matematicheskoĭ literatury khaldeev* ([*A Historical Sketch of the Mathematical Literature of the Chaldeans*], 1881) by the mathematician and historian of mathematics Mikhail Egorovich Vashchenko-Zakharchenko (1825–1912),[55] who worked in Kiev. The catalog dates back to 1881 or the very beginning of 1882. So the same author's subsequent book *Istoricheskiĭ ocherk matematicheskoĭ literatury indusov* ([*A Historical Sketch of the Mathematical Literature of the Hindus*], 1882) would have been published too late for inclusion in the catalog. The volume comprises an eloquent tribute to ancient Indian mathematicians:

> The Hindu vision of the external world was much wider and more sublime [gorazdo shire i velichestvennee] than that of the ancient Greeks. In their philosophy the Hindus succeeded in moving on from the examination of the natural bodies to the notions of the unlimited, boundless, formless and eternal; they started considering the world as something delusive and transient [nechto prevratnoe, prokhodîashchee]; the idea of shape and appearance [predstavlenie o forme i vide] gave place to the notions of matter and the divine spark [ponîatiîam o veshchestve i bozhestvennom nachale].[56]

DOI: 10.1057/9781137338280

One could read here with references to such authorities as Jean Sylvain Bailly (1736–1793) that Hindus had already been familiar with what was only later discovered by Ptolemy or even Newton. Moreover, we are told that there is a remarkable affinity between the most recent mathematical theories and those of the ancient Indians:

> Elliptic integrals and their reversed functions of various orders, spaces of various dimensions—are not these the various degrees of heaven in which Hindu gods are seated [Èllipticheskie integraly i obratnye im funktsii razlichnykh porîadkov, prostranstva razlichnykh izmereniï—ne sut' li èto razlichnye stepeni neba, v kotorykh vossedaîut indusskie bogi]?[57]

This proposed parallel between modern mathematics and Hindu gods is a variant on the general idea of the Oriental or Aryan renaissance, which gained considerable authority in the nineteenth century.[58] Modern mathematics is represented as the rediscovery of forgotten Indian truths, and thus, we may presume, as preparing the ground for a great mathematics of the future. The fact that this observation is not only of intellectual, but also of emotional value for the author, evoking gods amidst mathematical terms, seems to point to this same framework of great expectations for intellectual regeneration.

2

The Moscow mathematicians were familiar with these ideas, which found their way into the famous Brockhaus and Efron dictionary. Here is what one reads in the article *Mathematics* written by one of Bugaev's pupils:

> The nation that at the same time as the Greeks was at the head of the intellectual development of humanity was that of the Hindus. This position had, however, been occupied by them considerably earlier than by the Greeks. This one can see from the fact that at a time when the Greeks were still humble disciples of the Egyptians, the wisdom of Brahmans was glorified throughout the East. We even have some obscure evidence that some of the Greeks, namely Pythagoras and Democritus of Abdera, traveled to Hindustan to learn this wisdom.[59]

In this curiously contradictory account of the origins of mathematics, we find the combination of a traditional reference to the Greeks as European civilizers with a conviction that the latter themselves had been disciples of the wise Brahmans—a conviction prompting a poetical digression on

DOI: 10.1057/9781137338280

the noble nature of the "Hindu genius," unexpected in an encyclopedia article on mathematics:

> As the two great religious systems created by the Hindus, Brahmanism and Buddhism, demonstrate, the national features of the Hindu genius were a bent for philosophical contemplation and speculation, aimed at penetrating into the very essence of things and at grasping the boundless and incomprehensible, a yearning to construct such systems of philosophic-religious worldview that, representing a neat logical whole, would give answers to all the greatest and the most difficult questions and enigmas of the life of the macrocosm and the microcosm, of the universe and man. The Hindu meditations, aimed exclusively at the discovery of the inner relation between things, were little concerned with external transient forms. In this respect, the Hindu were very unlike the Greeks, for whom form was of such an importance. <...> losing much in precision and rigor [in comparison with the Greeks], they gain in depth and breadth.[60]

The comparison between the Greeks and the Hindus outlined above was an important component in the idea of the "Aryan renaissance." It was believed that the discovery of the Indian texts in the eighteenth century would bring about a renaissance similar to that of the fourteenth to sixteenth centuries, inspired by the discovery of the Greek and Latin texts. As Indian thought was considered, within this frame of reference, to have been much profounder than Greek, it was anticipated that the new renaissance would surpass the previous one, and would give rise to a new culture of incomparable beauty, wisdom, and humanity. The obvious flaw in this comparison between past and future renaissances was that there was no evidence that Indian texts had had any real impact on contemporaneous culture. Knowledge of Sanskrit had not become widespread.[61] But this did not hold back from entertaining such hopes, and instead contributed to them, by making the Aryan renaissance an easier idea to discuss: it was sufficient to refer to a spiritual affinity between the Aryans and the moderns.

By the second half of the nineteenth century, views of India as the source of all wisdom were already on the wane[62] and the year 1895, when Viktor Viktorovich Bobynin (1849–1919) published his article in the Brockhaus and Efron encyclopedia, was already a rather late date for this position. The history of mathematics was no exception to the general European drift towards skepticism regarding the Indian origins of European learning. This skepticism was certainly known to Moscow mathematicians, and Bobynin, who was a very competent scholar,[63]

DOI: 10.1057/9781137338280

was no doubt aware of it. Thus, in Bugaev's library there was a book by Ferdinand Hoefer entitled *Histoire des mathématiques: depuis leurs origines jusqu'au commencement du dix-neuvième siècle* (1874) which Bobynin could have hardly missed. Hoefer placed great emphasis on the need to "restrain the claims of Indian science to be of great antiquity [rabattre des prétentions de la science indienne à une haute antiquité]."[64] In particular, he called accounts of the Indian travels of Pythagoras "mere fables [de simple contes]."[65]

This position was also to be found in Russian sources. Petr Lavrovich Lavrov's (1823–1900) *Ocherk istorii fiziko-matematicheskikh nauk* ([*A Sketch of the History of Physical-Mathematical Sciences*], 1865–1866) offered a succinct survey of historical literature representing both positions: of those who believed in Indian origins of European culture, and of those who did not.[66] Lavrov was on the side of the latter. Accordingly, he denied the Oriental sources of Pythagorean doctrines.[67]

3

Bugaev's passion for India finds a parallel in his disciple Bobynin's praise of Indian mathematics.[68] The anti-Semitism of Bugaev's circle was thus apparently nourished by both contemporary politics and historical speculations.

At one point, Bugaev actually insists on the Russians being Aryans.[69] Furthermore, in Bobynin's passage on the "Hindu genius" quoted above, all the "Hindu" features—the depth of reflection, the lack of interest in form, the bent for creating all-encompassing worldviews—were also frequently cited by Slavophils as constituting the Russian national character, prompting the question as to whether Bugaev's view of the Russians as Aryans was connected to his ideas about the task of the Moscow Mathematical Society being to nurture mathematicians–philosophers[70] (which then suggested to Nekrasov the term "philosophic–mathematical school").

The proof that mathematicians held the Slavs and Aryans to be connected is provided by an obscure popular brochure, *Velikiĭ schet* [*The Great Count*], published in 1922 in Odessa by A. Filippov, and mostly concerning Indian mathematics. There was a section "Arithmetical fantasies of the Hindus" and one on "Arithmetical knowledge of Buddha." We read here about the high probability that Archimedes was a successor of the Hindu, of the poverty of fantasy of the Greeks, Romans and Jews

DOI: 10.1057/9781137338280

in comparison to the Hindus[71] etc. There is also the following passage on "The old Russian numbering":

> In old Russian arithmetical manuscripts one can find a system similar to that of the Hindus. It seems that the Russian Slavs, as well as the Hindus, were particularly disposed to the meditation over numbers. It is possible that owing to the same national trait our mathematicians—Chebyshev, Bugaev, Voronoï—came to be distinguished mainly due to their research on the theory of numbers.[72]

The Russians are thus presented as resembling the Hindus, with the Russian mathematicians, both ancient and modern, Bugaev among them, manifesting the same disposition to a particular kind of research.

Bugaev's pupil P. A. Nekrasov placed the enigmatic "sages of ancient Oriental culture" at the root of the great tradition leading up to the Moscow Mathematical Society, in a work dedicated to Bugaev's memory. The "sages" of ancient Greece are evoked as merely shadowing the "Oriental" ones.[73] Nekrasov, who in the same work quoted Bugaev on Russians being Aryans[74] and was hardly unaware of the aforementioned historical views, almost certainly meant Indian sages by "Oriental." If this supposition is correct, we have here an interesting instance of tracing the intellectual genealogy of the Moscow "school" back to Aryan wisdom.

Foundations of the Evolutionary Monadology (1893)

There is reason to believe that the philosophical tract mentioned above, the *Foundations of Evolutionary Monadology*, bears traces of Bugaev's passion for the echoes of Indian philosophy that he found in theosophists, and more particularly in Blavatsky.

In the commentaries to his correspondence with Alexander Blok, written in 1926, Belyĭ lets drop a remark that he "was pythagoreanized [pifagoreizirovan]" by his father who "hospitably accepted the ideas of Karma and reincarnation into his *Monadology* [gostepriimno prinimavshego idei Karmy i perevoploshcheniia v svoiu 'Monadologiiu']."[75] While a detailed analysis of the sources of the *Foundations of the Evolutionary Monadology* is beyond the task of the present book, we can point to the following theses of this work as corroborating Belyĭ's words: after the disintegration of complex Monads, the life of the central Monad of such complexes may continue into a different complex;[76] there is a

DOI: 10.1057/9781137338280

constant process of perfection of the Monads,[77] the ultimate aim being to "remove the difference between the Monad and the world" and to "achieve an infinite perfection and to stand over the world [dostignut' beskonechnogo sovershenstva i stat' nad mirom]."[78]

I suggest that the immediate source of these statements was Blavatsky. In her magnum opus, *The Secret Doctrine* (1888) Monad is one of the most frequent words, and the "evolution of the Monads" (or "Monadic evolution") one of the central subjects.[79] Blavatsky associates her discussion of Monads not with Leibniz, but with Pythagoras, whose doctrines "are Oriental to the backbone, and even Brâhmanical."[80]

"Monads" are also called here "Pilgrim-souls." Their number is limited, and they are in constant journey, or "spiritual evolution," "through various *states* of not only matter, but of self-consciousness and self-perception <...>."[81] Progressing thus, "*through individual merits and efforts*" (italics are by Blavatsky), each Monad, or Pilgrim, gradually rises to the state of the "One Unconditioned All," wherein, "plunged into the incomprehensible Absolute Being and Bliss of Paranirvâna, he reigns unconditionally."[82]

According to one of the paragraphs of Bugaev's tract, it is not philosophy that can account for the essence and origin of Monads, but "deep doctrines of the Unconditioned."[83] Given what has been said, it seems clear that his notion of the "Unconditioned," another favorite term of Blavatsky, was allied in Bugaev's mind with her writings. Bugaev was evidently trying to offer a metaphysical equivalent of the account of the "Monadic evolution" expounded in *The Secret Doctrine*. This is why Belyĭ wrote in his memoir that in his father's theory of the evolutionary monadology, Monads were not the same as those in Leibniz' *Monadologie*, and evolution was understood "not in the style of Spencer."[84]

Seen against this background, the most interesting feature of Bugaev's *Foundations* seems to be the following paragraph: "The simple Monads are never born, nor ever die."[85] This has to be compared with the Monads of Leibniz which are born "par des Fulgurations continuelles de la Divinité," and, being thus created, can be annihilated by the same power (§§ 47, 6). Within the context of Blavatsky's writings, which were allegedly based on the forgotten revelations of the Indian wisdom, this comparison takes us back to Bugaev's "idée fixe." Blavatsky is very hostile to the Judaeo–Christian belief in creation ex nihilo, which she constantly confronts with that of the "evolution of preexisting materials,"[86] preached by other religions, and most importantly by Indian ones.

DOI: 10.1057/9781137338280

It is impossible to say how Bugaev could have reconciled his passionate interest in theosophy with Christian Orthodoxy, or his political conservatism with religious views which were perhaps very far from Orthodoxy.[87] Four years after his father's death, Belyĭ delivered a severe condemnation of the "crude" and "utilitarian [khozĭaĭstvennyĭ]" Semitic idea of creation that had been poisoning Aryan peoples for centuries, and called to return to the native Aryan view of palingenesis, wherein nothing is born and nothing dies.[88]

Bearing this in mind, one is justified in considering Bugaev's attempt to rewrite Leibniz' *Monadologie* along theosophical lines to have been part of the history of the European nineteenth-century "Aryan renaissance."

Symbolism (1910)

When reflecting on the "Aryan" connotations of Bugaev's "mathematicians–philosophers" that were to be nurtured by the Moscow Mathematical Society, the next point to consider is regarding his son's collected articles, published under the title *Symbolism* (1910).

The importance of this thick volume of more than 600 pages lies primarily in its remarkable articles on the theory of verse. Roman Jakobson described them as decisive for his early studies on verse.[89] The whole atmosphere of enthusiasm for exact methods in criticism, characteristic of the second decade of the past century in Russia and culminating in the Formal school, which was to have a strong impact on the literary studies of the century, owed much to this strange book.

The articles on verse, however, only comprise a portion of *Symbolism*,[90] which is a curious mixture of philosophy, theosophy, and science. The works to be found here date back to different years, but the most important ones, as well as the huge commentary (about a third of the entire volume), were written specially for this book. One of the striking features of the commentary is its endless bibliographies, including amazingly diverse authors (from Poincaré to Blavatsky), which reminds one of the library of the author's father.

The intention of the book, according to the author, was to lay foundations for the future "symbolic worldview."[91] What the latter should comprise is not very clear, except that it must include everything, in the way mathematics was expected to do.

DOI: 10.1057/9781137338280

Accordingly, *Symbolism* perplexes with the bewildering range of subjects it is intended to cover, its obscurity of language (partly owing to an abundance of terms from different branches of knowledge), and the elusiveness of its general message.

The very word "symbolism" in the title is confusing. Although Belyĭ is one of the most famous Russian symbolists, the volume has little to do with literary symbolism. Trying to gloss the words "symbol" and "symbolism" as used in this book would require writing another, which would deal more with philosophy and science than with literature. For our purpose only one sense of this word is important, one that must have been crucial for Belyĭ at the time.

Symbolism was brought out by the publishing house "Musaget." Belyĭ himself had taken an active part in the organization of "Musaget," of which the chief editor was his close friend, the musical critic and ardent Germanophile Ėmiliĭ Medtner (1872–1936). One of the essential elements of the ideology behind this enterprise was the idea of the Aryan renaissance.[92] Medtner was an admirer of H. St. Chamberlain, whose *Arische Weltanschauung* (1905) was translated into Russian and published by "Musaget" (1913).[93] Medtner was preparing a study on Chamberlain. Although this project never materialized, some drafts have been preserved in his archives.[94] There have also been preserved almost 200 pages of his notes taken from Chamberlain's magnum opus *Die Grundlagen des neunzehnten Jahrhunderts*, which Medtner read at least twice during 1908–1910.[95] He was obviously fascinated by Chamberlain, who helped define the direction of Medtner's own activity, of which the essential part at that time was devoted to "Musaget."

In the list of books Belyĭ had read not long before his work on *Symbolism* we encounter Chamberlain's name.[96] Belyĭ could have hardly failed to come across the following dichotomy stressed by Chamberlain: while the Semites are born idolaters (hence strict laws against idolatry in Judaism),[97] it is the "all-embracing symbolism," "a symbolism which almost goes beyond our modern powers of conception" that defines the Aryan thought. Here is what he writes in the chapter "The Entrance of the Jews into the History of the West" of the *Foundations*:

> As his mind ripened, he [the Aryan] began more and more to realize, not merely that those mythological forms possessed existence in his brain only, had a meaning only for his special limited human spirit—in other words, were symbols of a something which the reason could not reach—but also that his whole life, the world that served him as a stage, the actors that

DOI: 10.1057/9781137338280

moved upon this stage, the thought that he thought, the love that intoxicated him, the duties he fulfilled, were to be regarded as mere symbols; he did not deny the reality of these things, but he denied that their significance was exhausted by the empirically perceptible. "On the standpoint of the highest reality, all empirical activity has no existence," say the sacred writings of the Hindoos <...>.[98]

It is known that at the time of preparing *Symbolism* for publication, Belyĭ was undergoing a dark period of overwhelming Judeophobia verging on persecution mania.[99] Such passages must have been of utmost importance for him. In view of the Aryan orientation of "Musaget" and the fact that *Symbolism* was one of its first ideological pronouncements, making this volume a manifesto of the new publishing house and the group of authors around it, we have every reason to believe that the "Aryan" component of the meaning of the word "symbolism" was very much in evidence for Belyĭ. "Symbolism" is very likely to have been perceived by him at that period as an Aryan weapon against Jewish idolatry, the chief expression of which latter was the destructive materialism of the modern age, which, so the argument, needed to be replaced by a new "symbolic worldview." Numerous references to Indian thought peppering *Symbolism*, and making its reading still more difficult, reveal Belyĭ's preoccupation with imaginary "Aryanism," and thus underline the general message of the work as stated explicitly at the end of the core article of *Symbolism* entitled *The Emblematic of Meaning*:

> <...> the theory of symbolism [if it succeeds in becoming "the base of a special kind of creative experience"] will be a new system among the existing systems of Hindu philosophy: those of Vedanta, Yoga, Mimansa, Sankhya, Vaisheshika, and others; most probably, the theory of symbolism will not be a theory at all, but a new religious-philosophical teaching anticipated by the whole course of the development of Western-European thought. <...> the great significance of Hindu philosophy is recognized by such an expert on this philosophy as Deussen.[100]

Still more telling is the evidence provided by Medtner's archives, where we find a document entitled *Symbolism* dating back to the years of Medtner's intense reading of Chamberlain and his organizing of "Musaget" (1908–1910). It is a short record of only nine pages filled with scattered quotations and remarks. The reference to *Grundlagen* emerges next to the very beginning.[101] Mention of the Hindus or the Semites is absent only on the first and on the last page. On all the others the words

DOI: 10.1057/9781137338280

"symbol" and "symbolism" are evoked in the constant company either of the Hindus, or—by contrast—of the Semites.[102]

One is naturally curious to know what factors brought about the universal character of Belyǐ's *Symbolism* with its articles on philosophy, psychology, literary theory etc. Was it a matter of taste and ambition (which had been fostered by his father)? Was this cultivated bent for versatility strengthened and confirmed by meditations on Aryan intellectual virtues? One of the features Chamberlain stressed as "Aryan" (and in this he was not at all original) was an inclination for a wide worldview in sharp contrast to the "Semitic" tendency for practical narrow-mindedness.[103] For a reader of Chamberlain and a contributor to an "Aryan" publishing house, it was only logical to pronounce on symbolism in the form of a book of all-embracing character.[104]

* * *

It is important to emphasize the presence of these ideas in the immediate environment of the Moscow mathematicians. Aspiring to create a universal outlook based on mathematics, Bugaev was no less interested in Indian thought than was his son when composing *Symbolism*. This probably tells us something about the origins of Bugaev's aspirations.

It should also be noted that in the course of the Soviet polemics over mathematics, the conception of "Aryan" mathematical genius was not forgotten. In his review of the book by Stepan Alexandrovich Bogomolov (1877–1965) *Ėvolutsiia geometricheskoĭ mysli* ([*The Evolution of Geometrical Thought*], 1928), Ernst Kolman (1892–1979) commented on the opposition drawn by the author between the Egyptians with their practical mind and primitive mathematical notions, and the Greeks, more apt to seek abstract knowledge without reference to practice. The commentary runs as follows: "So, owing to this racial peculiarity of the Aryans, loving the elevated [lĭubĭashchikh vozvyshennoe ariĭtsev] (in contrast to the brutish Hamites-Semites) there emerges the 'pure science, speculative geometry' <...>."[105] This interpretation appears to have been dictated by pure malevolence. However, Kolman, who was to become one of the most zealous persecutors of the Moscow "school," started as a Jewish nationalist, and never ceased to be one, despite his Marxism.[106] His was the view of a scholar well-read in the anti-Semitic literature of those times, one of the favorite clichés of which was the opposition between the "Aryans, loving the elevated" and the "brutish Hamites-Semites."[107] Against the

DOI: 10.1057/9781137338280

background of these aforementioned ideas, Kolman's critique, even if not correct (for Bogomolov's remarks on this subject are scanty) does not seem to be absurd.[108] This context as outlined here might prove useful for a better understanding of the Soviet campaign against the Moscow "school."

Notes

1   The best source on Bugaev remains the issue of *Matematicheskiĭ sbornik* dedicated to his memory: *MS* 25: 2 (1905). See also M.ÎA. Vygodskiĭ, "Matematika i ee deîateli v Moskovskom universitete vo vtoroĭ polovine XIX v.," *IMI* 1 (1948), 165–175; S.S. Demidov, "N.V. Bugaev i vozniknovenie moskovskoĭ shkoly teorii funktsiĭ deĭstvitel'nogo peremennogo," *IMI* 29 (1985), 115–123; V.A. Shaposhnikov, "Filosofskie vzglîady N.V. Bugaeva i russkaîa kul'tura kontsa XIX–nachala XX vv.," *IMI. Vtoraîa seriîa* 42: 7 (2002), 62–91. The documents from Bugaev's archives concerning his interest in the idea of a universal significance of mathematics are cited further on in this section.

2   ORK i R NB MGU, f. 41, op. 1, ed. khr. 252. The existence of this catalog was pointed out by A.V. Ulanova in her description of Bugaev's archives: "Arkhivnyĭ fond Nikolaîa Vasil'evicha Bugaeva v Otdele redkikh knig i rukopiseĭ Nauchnoĭ biblioteki MGU im. M.V. Lomonosova," in *Rukopisi. Redkie izdaniîa. Arkhivy. Iz fondov Otdela redkikh knig i rukopiseĭ Nauchnoĭ biblioteki MGU* (Moskva: Vodoleĭ Publishers, 2008), 54–55.

3   See Ilona Svetlikova, "Moskovskie pifagoreĭtsy," in S.N. Zenkin, ed., *Intellektual'nyĭ îazyk èpokhi: Istoriîa ideĭ, istoriîa slov [materialy mezhdunarodnoĭ konferentsii, organizovannoĭ RGGU pri uchastii NLO 16–17 fevralîa 2009 g.]* (Moskva: Novoe literaturnoe obozrenie, 2011), 124 n. 22.

4   His *Rakkurs k dnevniku* ([*A Retrospective Diary*], 1930; RGALI, f. 53, op. 1, ed. khr. 100) is in its considerable part a retrospective diary of a reader, containing references to what Belyĭ read in different periods of his life.

5   ORK i R NB MGU, f. 41, op. 1, ed. khr. 11. The inscription (of an obviously later date) on this copy-book erroneously attributes the source of these notes as the first edition of *Histoire des mathématiques* (1758).

6   J.F. Montucla, *Histoire des mathématiques*, T. 1 (Paris: Henri Agasse, 1799), 15 (italics are by Montucla). ORK i R NB MGU, f. 41, op. 1, ed. khr. 11, L. 1. Among Bugaev's notes, there is also one taken from the following passage: "Tout le monde connoît l'inscription fameuse, par laquelle il [Platon] défendoit l'entrée de son auditoire à ceux qui ignoroient la géométrie. Il disoit enfin que la Divinité s'en occupoient continuellement, entendant sans

doute par-là, que toutes les lois par lesquelles elle gouverne l'univers, sont des lois mathématiques <...>" (Montucla, *Histoire des mathématiques*, 1799, 164; ORK i R NB MGU, f. 41, op. 1, ed. khr. 11, L. 3). It is curious to notice that Plato's words to be found here in Bugaev's own hand had undoubtedly something to do with the latter's transformation in *Petersburg* into a god-like ruler continually preoccupied with geometry.

7   Montucla, *Histoire des mathématiques*, 1799, 17–18; ORK i R NB MGU, f. 41, op. 1, ed. khr. 11, L. 1 – 1 ob.

8   It features not only in his Montucla notes (ORK i R NB MGU, f. 41, op. 1, ed. khr. 11, L. 1 ob.), but, with more detail, in later notes taken from Kuno Fischer's *Leibniz und seine Schule* (1855; ORK i R NB MGU, f. 41, op. 1, ed. khr. 261, L. 54 ob. – 55). Cf. N.V. Bugaev, *Matematika kak orudie nauchnoe i pedagogicheskoe* (Moskva, 1869), 21.

9   This task was suggested by Herbartian psychology (see Section "Psychology").

10 Bugaev, *Matematika kak orudie nauchnoe i pedagogicheskoe*, 29–30. In reading this, one is reminded of the contrast between the godlike senator Apollon Apollonovich and his son, who resembles a frog, a contrast which Belyĭ emphasizes throughout *Petersburg*. As was pointed out, this motif derives from the tradition of physiognomy, in particular from J.G. Lavater: the frog therein represented the first stage of evolution which culminated in Apollo. See Arkadiĭ Blĭumbaum, "Apollon i lĭagushka," in *Na rubezhe dvukh stoletiĭ: Sbornik v chest' 60-letiĭa Aleksandra Vasil'evicha Lavrova* (Moskva: Novoe literaturnoe obozrenie, 2009), 70–85. Bearing in mind Belyĭ's deep interest in his father's works, this motif might have implied a subtle parody, the frog exemplifying the "animal instincts" to be overcome by the divine powers of mathematics "establishing the harmony and esthetical feeling."

11 N. Brashman, *O vliĭanii matematicheskikh nauk na razvitie umstvennykh sposobnosteĭ* (Moskva, 1841), 7.

12 It is useful to point out two interrelated sources of Brashman and Bugaev in this respect. Brashman's speech was intended as a refutation of William Hamilton's essay *On the Study of Mathematics as an Exercise for the Mind* (*Edinburgh Review* 62: 126 (1836), 409–455), in which mathematics was characterized as not only useless in educating mental faculties, but downright pernicious.

Hamilton's essay became very popular. In a note to its later edition the author proudly referred to the fact that this work had been translated into the main European languages, its ideas having found their way even into a Latin speech pronounced before "his Majesty of Sardinia." See William Hamilton, *Discussions on Philosophy and Literature, Education and University Reform*, 3rd edn (Edinburgh and London: William Blackwood and Sons, 1866), 260–261 n. Schopenhauer was one of those who approved of Hamilton's

DOI: 10.1057/9781137338280

treatment of mathematics; the reactions to this work were still to be found in the beginning of the twentieth century. See Florian Cajori, "A Review of Three Famous Attacks upon the Study of Mathematics as a Training of the Mind," *Popular Science* 80: 22 (1912), 360–365; on Schopenhauer's reading of Hamilton: Cajori, "A Review of Three Famous Attacks," 365–366.

In *An Examination of Sir William Hamilton's Philosophy* (1865) J. St. Mill devoted a chapter to severe criticism of Hamilton's views of mathematics. Mill's book was translated into Russian in 1868 and became fashionable among Russian intellectuals, which may have influenced Bugaev's speech *Matematika kak orudie nauchnoe i pedagogicheskoe,* dealing, among other things, with the same subject of educational value of mathematics, and delivered the following year: though very critical of Hamilton, Mill's own attitude to mathematics and mathematicians was reserved and patronizing (the Russian translation of *An Examination* was in Bugaev's library; ORK i R NB MGU, f. 41, op.1, ed. khr. 252, L. 72 ob.) Cf. V.A. Shaposhnikov, "K voprosu o filosofsko-metodologicheskikh interesakh N.D. Brashmana," *IMI. Vtoraia seriia* 48: 13 (2009), 71–74.

13   D'Alembert, "Géometre," in Diderot and D'Alembert, eds, *L'Encyclopédie ou Dictionnaire raisonné des sciences, des arts et des métiers,* T. 7 (Paris, 1757), 628.

14   A notable exception was Jósef Maria Hoëné-Wroński (1776–1853), an outstanding mathematician and a philosopher developing the most extravagant theories based on mathematics. In particular, he was engaged in meditations over social and political issues and wrote letters to various rulers, including the Tsar Nicholas I. Moscow mathematicians were aware of him. Bugaev's pupil V.V. Bobynin wrote a highly sympathetic account of his philosophy ("Filosofiia matematiki po ucheniiu Goëne Vronskogo," *Fiziko-matematicheskie nauki v ikh nastoiashchem i proshedshem. Zhurnal chistoĭ i prikladnoĭ matematiki, astronomii i fiziki, izdavaemyĭ V.V. Bobyninym* 2 (3): 1 (1886), 73–96; 2 (3): 3 (1886), 271–288; 2 (3): 4 (1886), 394–437; this study was published as a separate book (Moskva, 1894). In Bugaev's notes there is one on the book by Alexandre Erdan's *La France Mystique* (1855). Among the topics of the book, such as the Mormons and "the sorcerers of the latest times [kolduny noveĭshego vremeni]," there emerges Wroński's name (ORK i R NB MGU, f. 41, op. 1, ed. khr. 256, L. 2). The note is an unmarked (and not accurate) quotation from a review of the second edition of the book (1858) published in *OZ* 124 (1859), 69. Having cited the review, Bugaev adds: "to buy [kupit']." Belyĭ tried reading Wroński after conversations with another disciple of his father, Florenskiĭ. See Andreĭ Belyĭ, *Nachalo veka* (Moskva: Khudozhestvennaia literatura, 1990), 300.

15   This example is taken from an author read and reread by Belyĭ: Karl von Eckartshausen, *Nauka chisl* [Zahlenlehre der Natur], Ch. I (St Peterburg, 1815), 58 (Belyĭ, *Rakkurs k dnevniku,* L. 29, 48 ob.)

DOI: 10.1057/9781137338280

It should be noted that traces of a similar attitude were to be found in contemporary presentations of theory of probability. The latter was described, firstly, as a discipline of wide relevance. Thus, the reader of Pushkin's *Sovremennik* knew that theory of probability was of vital import as a remedy for wishful thinking and misleading hopes. See [P.B.] Kozlovskiĭ, "O nadezhde," *Sovremennik*, 3 (1836), 23–47 (I thank Vera Mil'china for this reference). Secondly, theory of probability was characterized as providing one with an insight into the laws of divine providence (Brashman, *O vliĩanii matematicheskikh nauk na razvitie umstvennykh sposobnosteĭ*, 29; cf. Chapter 2, Section "The divine order.")

16 He had more than 30 books on this subject. Apart from the special section on the history of mathematics (ORK i R NB MGU, f. 41, op. 1, ed. khr. 252, L. 8–9), there were books on the history of particular areas of mathematics related to other sections (L. 1, 11, 30 ob., and 50 ob.)

17 Johann Heinrich Moritz Poppe, *Geschichte der Mathematik seit der älteste bis auf die neueste Zeit* (Tübingen: C.F. Osiander, 1828), 1; cf. Ferdinand Hoefer, *Histoire des mathématiques: depuis leurs origines jusqu'au commencement du dix-neuvième siècle* (Paris: Librairie Hachette et Cⁱᵉ, 1874), 3. All the examples related to the history of mathematics are taken from books which were in Bugaev's library, with some specified exceptions, mainly concerning historical works published after the catalog of the library was drawn up.

18 Hoefer, *Histoire des mathématiques*, 92. Cf. Moritz Cantor, *Mathematische Beiträge zum Kulturleben der Völker* (Halle: H.W. Schmidt, 1863), 80; Hermann Hankel, *Zur Geschichte der Mathematik in Altertum und Mittelalter* (Leipzig: B.G. Teubner, 1874), 110.

19 Some useful indications concerning Bugaev's early positivism are to be found in Shaposhnikov, "Filosofskie vzgliady N.V. Bugaeva," 66–68.

20 ORK i R NB MGU, f. 41, op.1, ed. khr. 252, L. 76 ob. A considerable amount of books related to positive philosophy (Comte, Mill, Spencer, Lewes and Russian works on them) were held in the section of Bugaev's library on the history of philosophy (ORK i R NB MGU, f. 41, op.1, ed. khr. 252, L. 72–73 ob.)

21 See P.A. Nekrasov, "Filosofiia i logika nauki o massovykh proĩavleniĩakh chelovecheskoĭ deĩatel'nosti (Peresmotr osnovaniĭ sotsial'noĭ fiziki Quetelet)," *MS* 23: 3 (1902), 466–474; P.A. Nekrasov, "Moskovskaĩa filosofsko-matematicheskaĩa shkola i ee osnovateli," *MS* 25: 1 (1904), 11–14. One of Nekrasov's important references was the former president of the Moscow Mathematical Society (1886–1891) Vasiliĭ Iakovlevich Tsinger (1836–1907; *MS* 25: 1 (1904), 13 n.). Cf. Demidov, "N.V. Bugaev i vozniknovenie moskovskoĭ shkoly teorii funktsiĭ deĭstvitel'nogo peremennogo," 117.

22 This book was in Bugaev's library: ORK i R NB MGU, f. 41, op. 1, ed. khr. 252, L. 63.

DOI: 10.1057/9781137338280

23   Between reading Whewell in 1899 (*Rakkurs k dnevniku*, L. 2) and writing
     Petersburg Belyĭ had studied a great deal of occult literature treating of
     astrology. It is, however, curious to note that the cold and dry nature ascribed
     to Saturn by astrological tradition and used by Belyĭ in his representation
     of Apollon Apollonovich (in his son's dream the latter is time and Saturn
     himself) must have been known to him since his reading of Whewell
     (*Istoriia induktivnykh nauk ot drevneĭshego i do nastoĭashchego vremeni*, T. I
     (Sankt-Peterburg, 1867), 180, 372). Likewise, some occult doctrines could
     have been first found in critical accounts of the history of sciences, and then
     re-discovered in occult sources. This might have made their adoption easier:
     one "revised" what one had already known.

24   ORK i R NB MGU, f. 41, op. 1, ed. khr. 252, L. 78–78 ob.

25   Reading P.L. Lavrov, Bugaev was aware of this role of psychology as early as
     1860 (ORK i R NB MGU, f. 41, op.1, ed. khr. 256, L. 3). Lavrov stated that the
     dominant role in the hierarchy of sciences had passed on from mathematics
     to natural sciences and psychology primarily ("Sovremennoe sostoĭanie
     psikhologii," *OZ* 129 (1860), 49–50).

26   See I.ĬU. Svetlikova, *Istoki russkogo formalizma: Traditsiia psikhologizma i
     russkaia formal'naia shkola* (Moskva: Novoe literaturnoe obozrenie, 2005),
     15–40.

27   ORK i R NB MGU, f. 41, op.1, ed. khr. 252, L. 74–75 ob.

28   They are held in the folder entitled "Materials on psychology and other
     scientific branches" (ORK i R NB MGU, f. 41, op.1, ed. khr. 261).

29   See Svetlikova, *Istoki russkogo formalizma*, 23–28, 114–123.

30   N.I. Shishkin, "O determinizme v sviazi s matematicheskoĭ psikhologieĭ,"
     VFiP 8: 4 (1891), 110–128. On Shishkin, see Shaposhnikov, "Filosofskie
     vzgliady N.V. Bugaeva," 88 n. 28.

31   ORK i R NB MGU, f. 41, ed. khr. 256, L. 2 ob.

32   *OZ* 130 (1860), 1–10. Bugaev used to read this journal when young (ORK i R
     NB MGU, f. 41, op. 1, ed. khr. 256, L. 2, 2 ob., 3).

33   After having drawn up the catalog of his library, Bugaev bought this book.
     Belyĭ read his father's copy; see Andreĭ Belyĭ and Ivanov-Razumnik, *Perepiska*
     (Sankt-Peterburg: Atheneum; Feniks, 1998), 491.

34   Andreĭ Belyĭ, *Na rubezhe dvukh stoletiĭ* (Moskva: Khudozhestvennaia
     literatura, 1989), 231.

35   S.M. Solov'ev, *Vladimir Solov'ev: Zhizn' i tvorcheskaia èvoliutsiia* (Moskva:
     Respublika, 1997), 43.

36   ORK i R NB MGU, f. 41, op. 1, ed. khr. 87, L. 65. In Bugaev's library there
     were such musts for spiritualists as Allan Cardec and William Crookes
     (ORK i R NB MGU, f. 41, op. 1, ed. khr. 252, L. 78–78 ob). In the *Christened
     Chinaman* (1921) Belyĭ depicts Mikhail Vasil'evich Letaev, whose prototype
     was his father, as a believer in transmigration of the soul. See Andreĭ Belyĭ,

DOI: 10.1057/9781137338280

Sobranie sochineniĭ (Moskva: Respublika, 1997), 243; for Bugaev's reading of Blavatsky, see Section *"Foundations of the Evolutionary Mondadology."* In the context of contemporary psychology this belief acted as a corollary to observing psychic phenomena.

37   S. Solov'ev, *Vospominaniia* (Moskva: Novoe literaturnoe obozrenie, 2003), 171.

38   ORK i R NB MGU, f. 41, op.1, ed. khr. 256, L. 2 ob. – 3.

39   ORK i R NB MGU, f. 41, op.1, ed. khr. 256, L. 2 – 2 ob.

40   ORK i R NB MGU, f. 41, op.1, ed. khr. 333, L. 107. Nekrasov must have meant the fact that the Grand Duke helped to increase the subsidy received by the Society. See the minutes of the sitting of the society of April 26, 1894 (*MS* 17: 4 (1895), 834); see also Paul Buckingham, "Mathematics as a Tool for Economic and Cultural Development: The Philosophical Views of the Leaders of the Moscow Mathematical Society, 1867–1905," *Michigan Academician* 31:1 (1999), 34–35.

41   The appointment of the Grand Duke Sergeĭ Aleksandrovich to the post of general-governor was hailed by conservatives as the beginning of a new period of Russian history, whereby Moscow, the "true Russian capital," would eclipse the cultural influence of St. Petersburg. See D.B. Grishin, *Tragicheskaia sud'ba velikogo kniazia* (Moskva: Veche, 2006), 150. In the *Christened Chinaman*, devoted to his childhood, that is to the time close to that appointment, Belyĭ makes Mikhail Vasil'evich Letaev (i.e. his father) emphatically voice the same views: "Moscow is <…> the natural, Russian center of ours, of all the intellectual, literary, social life…"; Petersburg is "the Germans [nemchura]" (Belyĭ, *Sobranie sochineniĭ*, 205). See also L.K. Lakhtin, "Nikolaĭ Vasil'evich Bugaev (Biograficheskiĭ ocherk)," *MS* 25: 2 (1905), 258.

42   Nekrasov, "Moskovskaia filosofsko-matematicheskaia shkola," 248; Vygodskiĭ, "Matematika i ee deiateli," 166.

43   ORK i R NB MGU, f. 41, op.1, ed. khr. 164, L. 6 – 6 ob.

44   ORK i R NB MGU, f. 41, op.1, ed. khr. 274.

45   A brief mention of the Grand Duke in Belyĭ's memoir is significant. In one and the same passage Belyĭ makes his father name the Tsar Nicholas II "mal'chishka [a silly boy]" and the Grand Duke "pederast" (*Na rubezhe dvukh stoletiĭ*, 61). Whether it was invented or not (even among many conservatives the reputation of both the Tsar and the Grand Duke was pretty bad), the phrase is suggestive: Bugaev's circle had been close to some of the most notorious reactionaries, and it was only natural on the part of Belyĭ writing his memoirs in the Soviet times to try to distance his father from such connections.

46   P.A. Nekrasov, *Gosudarstvo i Akademiia* (Moskva, 1905), 5 n.

47   It dates April 1, 1893 (ORK i R NB MGU, f. 41, op.1, ed. khr. 333, L. 117).

DOI: 10.1057/9781137338280

48   In Bugaev's archives there is also a letter of invitation from Alekseĭ
     Semenovich Shmakov (1852–1916), the lawyer, who later became one of the
     most notorious and prolific Russian anti-Semitic writers. From one of his
     books, *Svoboda i evrei* ([*Freedom and the Jews*], 1906) Belyĭ drew inspiration
     while working on *Petersburg*. In the letter dated June 19, 1900, Shmakov
     invited Bugaev to the sitting of the board of administration of the Moscow
     State Credit Society (ORK i R NB MGU, f. 41, op. 1, ed. khr. 357). Bugaev
     was there and reported to his wife that "they turned out to be scoundrels
     [okazalis' negodiaĭami]." The ground for this remark was, however, not
     ideological, but financial (ORK i R NB MGU, f. 41, op. 1, ed. khr. 311,
     L. 150 ob.)

49   In his work dedicated to Bugaev's memory, Nekrasov easily interweaves
     monadology with his autocratic monarchism: he writes of "every complex
     monad" as presupposing "a uniting autocratic (souvereign) principle
     [samoderzhavnoe (suverennoe) nachalo]," the Russian autocratic state
     [samoderzhavie] being thus presented as an organic part of this general
     metaphysical whole (Nekrasov, "Moskovskaia filosofsko-matematicheskaia
     shkola," 105).

50   Lakhtin, "Nikolaĭ Vasil'evich Bugaev," 258.

51   It is dated April 1902 (ORK i R NB MGU, f. 41, op. 1, ed. khr. 236, L. 1). In
     Belyĭ's *Christened Chinaman* "Austrophil and liberal" are synonyms and swear
     words with Letaev the elder (Belyĭ, *Sobranie sochineniĭ*, 171–173).

52   ÎA. Veĭnberg, *Nikolaĭ Kopernik i ego uchenie* (S.-Peterburg, 1873); J. Bertrand,
     Copernic et ses travaux (Paris, 1864); C. Flammarion, *Vie de Copernic* (Paris,
     1872); Studnička, *Mikuláš Koperník* (Praga, 1873) (ORK i R NB MGU, f. 41,
     op. 1, ed. khr. 252, L. 8 – 8 ob.) There was a heated argument at that time
     whether Copernicus was of German or Slavic origins. For a detailed
     discussion of this question, see Veĭnberg, *Nikolaĭ Kopernik*, 49–73.

53   Solov'ev, *Vospominaniia*, 171. One of the first and most prominent Russian
     theosophists, Anna Sergeevna Goncharova (1855 –?) was Bugaev's friend
     (Belyĭ, *Na rubezhe dvukh stoletiĭ*, 249–250).

54   Belyĭ, *Na rubezhe dvukh stoletiĭ*, 58.

55   He was Bugaev's acquaintance (Belyĭ, *Sobranie socineniĭ*, 172).

56   M.E. Vashchenko-Zakharchenko, *Istoricheskiĭ ocherk matematicheskoĭ
     literatury indusov* (Kiev, 1882), 1.

57   ibid., 76.

58   See Raymond Schwab, *La Renaissance orientale* (Paris: Payot, 1950); Roger-Pol
     Droit, *Le Culte du néant: Les philosophes et le Bouddha* (Paris: Seuil, 2004),
     19–21; id., *L'Oubli de l'Inde: Une amnésie philosophique* (Paris: Seuil, 2004),
     107–133.

59   V. Bobynin, "Matematika," in F.A. Brokhaus and I.A. Efron, eds,
     Ėntsiklopedicheskiĭ slovar', T. XVIIIᴬ (S.-Peterburg, 1896), 789.

DOI: 10.1057/9781137338280

60 ibid., 789–790. Bobynin was fascinated by Indian mathematics for years. His earlier works contain similar reflections, see "Periody, napravleniia i shkoly v razvitii nauk matematicheskikh," *Fiziko-matematicheskie nauki v ikh nastoiashchem i proshedshem* 2 (3): 1 (1886), 7–8; "Filosofskoe, nauchnoe i pedagogicheskoe znachenie istorii matematiki," *Fiziko-matematicheskie nauki v ikh nastoiashchem i proshedshem* 1: 2 (1885), 103–104.

61 It is telling that Schopenhauer, one of whose favorite books was the *Upanishads*, and who had a remarkable linguistic talent, did not know Sanskrit (Droit, *L'Oubli de l'Inde*, 171).

62 Theosophy was an important exception to this tendency of "forgetting India" in the latter half of the nineteenth century. For a detailed analysis of the change of attitude to India at that time, see Droit, *L'Oubli de l'Inde*.

63 On his works on the history of mathematics, see T.A. Tokareva, "Istoriia matematiki v Rossii: rozhdenie distsipliny," *IMI. Vtoraia seriia* 44: 9 (2005), 221–232; V.P. Zubov, "Bobynin i ego trudy po istorii matematiki," in *Trudy instituta istorii estestvoznaniia i tekhniki*, T. 15 (Moskva, 1956), 277–322.

64 Hoefer, *Histoire des mathématiques*, 50.

65 ibid., 88.

66 P.L. Lavrov, *Ocherk istorii fiziko-matematicheskikh nauk* ([Sankt-Peterburg, 1865–1866]), 50–52. Lavrov was for some time an author of *Otechestvennye zapiski*, and, as the latter's notes show, one of the favorite with Bugaev (ORK i R NB MGU, f. 41, op.1, ed. khr. 256, L. 2, 3, 5).

67 Lavrov, *Ocherk istorii fiziko-matematicheskikh nauk*, 87 n. 50.

68 Bugaev's libretto *Buddha* may have had something to do with the interest towards Indian mathematics: Buddha's mathematical skills were widely known in this context.

69 *Trudy Vysochaishe uchrezhdennoi komissii po voprosu ob uluchsheniiakh v srednei obshcheobrazovatel'noi shkole*, Vyp. VI (S.-Peterburg., 1900), 77. These words turn up in the course of the discussion around the classical system of education. In the catalog of Bugaev's library there is the Russian translation of Karl Schmidt's (1819–1864) fundamental *Geschichte der Pädagogik* (1860–1862; Istoriia pedagogiki, 1877–1881; ORK i R NB MGU, f. 41, op.1, ed. khr. 252, L. 77–77 ob.) In the first volume of the fourth edition of this book, devoted to the pre-Christian era and, among other things, to Hindu education, Bugaev could find the corroboration of his views. The character of the Hindus was described by Schmidt as in Bobynin's article, and the Slaves were stated to be Aryans (*Istoriia pedagogiki*, T. 1 (Moskva: Izd. K.T. Soldatenkova, 1890), 346, 292; on the role of the Hindus in the history of mathematics, 348).

On the history of the Aryan myth in Russia, see Marlène Laruelle, *Mythe aryen et rêve impérial dans la Russie du XIXe siècle* (Paris: CNRS, 2005).

DOI: 10.1057/9781137338280

70   *Rech' prezidenta Moskovskogo Matematicheskogo Obshchestva N.V. Bugaeva*
      (Moskva, [1896]), 6–7.
71   A. Filippov, *Velikiĭ schet* (Odessa: Vseukrainskoe gosudarstvennoe
      izdatel'stvo, 1922), 9.
72   ibid., 12–13.
73   Nekrasov, "Moskovskaia filosofsko-matematicheskaia shkola," 7.
74   ibid., 240.
75   Andreĭ Belyĭ and Aleksandr Blok, *Perepiska, 1903–1919* (Moskva: Progress-
      Pleiada, 2001), 71.
76   N.V. Bugaev, *Osnovy èvolutsionnoĭ monadologii* (Moskva, 1893), 11.
77   ibid., 11.
78   ibid., 13.
79   It comes out clearly in *Index* to the work: H.P. Blavatsky, *The Secret Doctrine:*
      The Synthesis of Science, Religion, and Philosophy, 3rd ed., *Index to Vols. I. and II*
      (London – New York – Benares: The Theosophical Publishing Society, 1895),
      178–180. See, in particular, Vol. I (1893), 192–193. It should be noted that
      Bugaev read English.
80   H.P. Blavatsky, *The Secret Doctrine: The Synthesis of Science, Religion, and*
      Philosophy, 3rd edn, *Vol. III* (London – New York – Benares – Chicago: The
      Theosophical Publishing Society; Theosophical Book Concern, 1897), 13.
81   Blavatsky, *The Secret Doctrine*, Vol. I (London – New York – Madras: The
      Theosophical Publishing Society, 1893), 198. Cf. in Bugaev: "The result of this
      development [of monads] will be a different, more elevated understanding
      of the world, a deeper feeling <...> [drugoe bolee vysshee ponimanie mira,
      bolee glubokoe chuvstvo]" (*Osnovy èvolutsionnoĭ monadologii*, 12).
82   Blavatsky, *The Secret Doctrine*, Vol. I, 288–289. In Bugaev Monads are engaged
      in the same movement of "rising up [podniatiia]" through their "efforts"
      (*Osnovy èvolutsionnoĭ monadologii*, 8). Cf. also in Blavatsky's earlier work *Isis*
      Unveiled (1877): "<...> Buddha, as a generic name, is the anthropomorphized
      monad of Pythagoras" (*Isis Unveiled: A Master-Key to the Mysteries of Ancient*
      and Modern Science and Theology, Vol. I (Theosophy Trust, 2006), 261).
83   Bugaev, *Osnovy èvolutsionnoĭ monadologii*, 17. Cf. Bugaev's claim that
      "monadological worldview" is conformable with "all the deepest doctrines of
      the Unconditioned" (*Osnovy èvolutsionnoĭ monadologii*, 18).
84   Belyĭ, *Na rubezhe dvukh stoletiĭ*, 172. Belyĭ's poem to the memory of his
      father, in which the latter is made to reiterate some of the essential points
      of his monadology, yields another indirect proof of our interpretation.
      "The floating monads" who are to become "the world" or to "stand over the
      world" ("My stanem—mir. Nad mirom vstanem my"), which is almost a
      word by word quotation from the *Foundations of the Evolutionary Monadology*,
      are moving in the "eon waves of the swashing times [v èonnykh volnakh
      pleshchushchikh vremen]." "Eons" here are derived not from Gnostics,

DOI: 10.1057/9781137338280

but from Blavatsky, in whose description of evolution the "eons of time" is a frequent synonym of "period." See Andreĭ Belyĭ, *Stikhotvoreniia i poemy*, T. 2 (Moskva – Sankt-Peterburg: Akademicheskiĭ Proekt; Progress-Pleîada, 2006), 410; theosophy is mentioned in the commentaries to this poem by A.V. Lavrov and J. Malmstad (608).

85 Bugaev, *Osnovy èvolutsionnoĭ monadologii*, 18.

86 Blavatsky, *Secret Doctrine*, Vol. I, 253.

87 Being an auditor at the Moscow University, one of V.A. Gringmut's interests was Sanskrit. See *Vladimir Andreevich Gringmut. Ocherk ego zhizni i deîateĺnosti* (Moskva, 1913), 15. This suggests an interest in India, which could have had something to do with Bugaev's idea of sending his tract to Gringmut. The latter's response is purely formal and gives no clues.

88 Taciturno [Belyĭ], "Palingenez," *Pereval* 6 (1907), 45–49.

89 Roman Jakobson, "Retrospect," in his *Selected Writings*, Vol. V (The Hague – Paris – New-York: Mouton Publishers, 1979), 569–570.

90 Andreĭ Belyĭ, *Simvolizm* (Moskva: Musaget, 1910), 231–428.

91 ibid., I, 51.

92 Omry Ronen, *Shram. Vtoraîa kniga iz goroda Ann* (Sankt-Peterburg: zhurnal "Zvezda," 2007), 70; cf. Mikhail Bezrodnyĭ, "Iz istorii russkogo germanofiĺstva: izdateĺstvo 'Musaget'," *Issledovaniîa po istorii russkoĭ mysli. Ezhegodnik za 1999 god* (Moskva: OGI, 1999), 166–167.

93 Magnus Ljunggren, *Russkiĭ Mefistofeĺ. Zhizn' i tvorchestvo Èmiliîa Metnera* (Sankt-Peterburg: Akademicheskiĭ proekt, 2001), 27.

94 RGB, f. 167, k.18, ed. khr. 9. This document bears the title: "'Chamberlain. The Culture. The Race. The Civilization. Historical Ideas'—materials for a study of the worldview of H. St. Chamberlain. 1903–1913." It is as long as 114 pages most of which are taken by notes from Chamberlain's works.

95 There is a 100 page conspectus of Chamblerlain's *Grundlagen* made in April–August of 1909 (RGB, f. 167, k.18, ed. khr.12). Another document of almost 100 pages entitled "A Diary" and dated to 1908–1910 is mostly taken up by notes from the same source (RGB, f. 167, k. 22, ed. khr. 13).

96 Belyĭ, *Rakkurs k dnevniku*, L. 45.

97 Cf. Renan's views on the matter pointed out in Maurice Olender, *Les Langues du Paradis. Aryens et sémites: un couple providential* (Paris: Seuil, 2002), 129–130. While preparing *Symbolism* Belyĭ read Swedenborg (in his *Retrospective Diary* he names *Arcana coelestia* and *L'apocalypse expliqué*; *Rakkurs k dnevniku*, L. 48 ob.), who held the same views: "The Jews were more idolatrous than other nations, and regarded their external rituals as constituting divine worship <…>. From the historical and prophetical parts of the Word, it is obvious that the Jews were prone to the worship of idols, and from the internal sense it is manifest that they were continually in idolatry <…>. Idolatry was so severely interdicted to the Jewish nation,

DOI: 10.1057/9781137338280

because the adoration of other gods and of images would have destroyed the representative of the church with them <...>" (*Index to Swedenborg's Arcana Coelestia or Heavenly Mysteries Contained in the Holy Scripture*, Vol. I. *A to L* (London: The Swedenborg Society, [1853]), 357).

98   Houston Stewart Chamberlain, *The Foundations of the Nineteenth Century*, trans. John Lees, 2nd edn, Vol I (London – New York: The Bodley Head; John Lane Company, 1912), 437. In Medtner's diary there is the following note: "The abstract materialism of the Jews. The Jews and the Semites in general are the only idolaters, that is the worshipers of the idol, and not of the idea symbolized by the idol [edinstvennye idolopoklonniki, t.e. poklonniki idola, a ne simvoliziruemoĭ im idei]. Prophets had fought against this, and exterminated this trait; the very notion of idol is purely Semitic; it did not occur to the Aryans to persecute idolatry (together with the visual arts)" (RGB, f. 167, k. 22, ed. khr. 13, L. 11 ob.; cf. Chamberlain's *Foundations*, 225).

99   See, primarily, Mikhail Bezrodnyĭ, "O 'iudoboĭazni' Andreĭa Belogo," *Novoe literaturnoe obozrenie* 28 (1997), 100–125.

100   Belyĭ, *Simvolizm*, 140. The massive presence of Kantianism in *Symbolism* points to the same direction. Belyĭ actually uses some terms from Indian philosophy (or rather theosophy) as equivalents of Kantian terms. The representation of Kant as a reviver of Indian wisdom was well known in the circle of "Musaget." It was there that Paul Deussen's *Vedânta und Platonismus im Lichte der Kantischen Philosophie* (1904) was translated (1911). In his notes Medtner mentions Chamberlain's *Immanuel Kant* (1905), which was one of the sources for this representation (RGB, f. 167, k. 18, ed. khr. 12, L. 1). For racial elements of Kantianism in Belyĭ and "Musaget," see Ilona Svetlikova, "Kant-semit i Kant-ariets u Belogo," *Novoe literaturnoe obozrenie* 93: 5 (2008), 62–98.

101   RGB, f. 167, k. 17, ed. khr. 8, L. 3.

102   Here is an example: "Since one cannot dispense with the myth in philosophy, its open symbolical recognition, as in the Hindus, is better than its sham banishment and secret emergence under the guise of despotic dogmatic superstitious <?> authority" (RGB, f. 167, k. 17, ed. khr. 8, L.5). The latter part of this phrase refers to the Semites, with the undercurrent theme of their being inborn idolaters.

103   "All-embracing" is one of the key characteristics of the Aryans in Chamberlain. Thus, Kant's "all-embracing mind possesses the whole world" (*Foundations*, 407), whereas "narrow-mindedness" is constantly brought in as a typical feature of the Semites.

104   See also Appendix 2.

       Symbolism is too uneven, too large, embraces too many subjects to have been understood and influential as a whole. Its Aryan message was

DOI: 10.1057/9781137338280

probably lost on most of its readers. It was the parts devoted to the theory of verse that had a real impact. And it was the Russian Formal School, consisting mostly of Jews, that developed Belyĭ's ideas and thus perpetuated them. This may seem paradoxical but was only characteristic of the time. See Vladimir (Zeev) Zhabotinskiĭ, "Strannoe iavlenie," in his *Izbrannoe* (Biblioteka-Aliia, 1990), 153–158.

105   È. Kol'man, "Bozhestvennaia èvoliutsiia geometricheskoĭ mysli," *Estestvoznanie i marksizm* 1 (1929), 157.

106   On Kolman, see, primarily, Eugene Seneta, "Mathematics, Religion, and Marxism in the Soviet Union in the 1930s," *Historia Mathematica* 31: 3 (2004), 362–364, 365–367.

107   S.A. Bogomolov, *Èvolutsiia geometricheskoĭ mysli* (Leningrad: Nachatki znaniĭ, 1928), 6–7, 10–11.

108   Cf. Chapter 4, Section "P.A. Florenskiĭ"; Chapter 5, Section "The secret radicalism of the 'school.'"

DOI: 10.1057/9781137338280

2

The Moscow "School": P. A. Nekrasov

Abstract: *In his work* The Moscow Philosophic-Mathematical School and Its Founders *(1904), P. A. Nekrasov presented mathematics as fundamental to the whole body of knowledge, including philosophy, theology and political theory.*

Svetlikova, Ilona. *The Moscow Pythagoreans: Mathematics, Mysticism, and Anti-Semitism in Russian Symbolism.* New York: Palgrave Macmillan, 2013. DOI: 10.1057/9781137338280.

DOI: 10.1057/9781137338280

Introduction

P. A. Nekrasov's (1853–1924) preoccupation with the idea of mathematics as the fundamental discipline and as a mighty tool for shaping a better culture and society owed much to Bugaev.[1] This statement could be extended to the whole circle of the Moscow "school." Obviously, not everybody who is to be mentioned further lacked common sense to such an extent as Nekrasov did. Nevertheless, his writings, in their exaggerating and distorting way, mirror tendencies that had originated with Bugaev and were then disseminated among his pupils.

1

However bizarre Nekrasov's ideas might seem, they cannot be dismissed as those of a marginal crackpot. According to historians of mathematics, he was an eminent mathematician.[2] What is even more significant, as we are concerned with the history of ideology, he was an important functionary: rector of Moscow University (1893–1898), warden of Moscow School District (1898–1905), promoted to be a member of the Council of the Ministry of Public Education in St. Petersburg.[3] This means that, contrary to the impression that Nekrasov's books convey, these are not the fancies of a lonely thinker divorced from practical life. This particular thinker had some power to realize his fancies. Even if he ultimately failed, his efforts were accompanied by a flow of books, booklets, articles and public discussions. What seems to us (and to Nekrasov's opponents) so fantastical was in fact less so, in the sense that it was at least discussed as a real possibility. Such possibilities are sometimes more indicative of a particular historical situation than actual events, the latter being more dependent on chance than unrealized projects carefully thought over (even if in such a strange manner as that of Nekrasov's). Thus it is important to emphasize, for it perceptibly changes our reading of the works to be discussed further, that the political ideas of the Moscow philosophic–mathematical school were part of contemporaneous political life, and it is as such that we have to regard them. They can by no means, however great the temptation is, be discarded as mere curiosa reflecting nothing but the troubled mind of some eccentric intellectuals.

2

Nekrasov's book *Moskovskaia filosofsko-matematicheskaia shkola i ee osnovateli* ([*The Moscow Philosophic-Mathematical School and Its Founders*],

DOI: 10.1057/9781137338280

1904) commemorating Bugaev, who had died the previous year, immediately immerses us in an atmosphere of universal knowledge. However perplexed we might be while reading *Symbolism*, written by Bugaev's son, this work by Bugaev's disciple leaves us in a state of utter bewilderment.

The book, to which we owe the term the "Moscow philosophic-mathematical school" and a somewhat ill-founded belief that this "school" was more than simply the name of an ideological trend, was published as an issue of a very respectable organ of the Moscow Mathematical Society, *Matematicheskiĭ sbornik*. Moreover, it was written by the president of this society, for Nekrasov replaced his teacher Bugaev in this position. In view of this, the *Moscow Philosophic-Mathematical School* must be regarded as a statement of the new president about the aims and ideals of the Moscow Mathematical Society.

This provides us with the necessary background. But rather than diminishing our perplexity by presenting circumstances rendering what we are reading less bizarre, this background contributes to our perplexity as a reminder that we are dealing with institutionalized realities. We are confronted with a project formulated in a language so strikingly awkward and obscure as to make some historians suppose that Nekrasov had developed a mental illness[4]; this project is highly unusual for that time, and contrary to all reasonable expectations it existed not in the mind of a marginal visionary, or an eccentric professional, like Wroński, but of the president of the Moscow Mathematical Society.

What do we find in this book? I shall cite from the table of contents:

1. Introductory words about the connection of Nikolaĭ Vasil'evich [Bugaev] to the activity of the Moscow Mathematical Society.—2. The Union [Soĭuz] of the founders of the Moscow Mathematical Society and its aim.—3. The ancient principles of exact knowledge and the principles of the thinkers of Western Europe.—4. The character of deeply penetrating human logos [in Russian it is no less obscure: kharakter gluboko pronikaĭushchego chelovecheskogo logosa].—5. On appraising the logos of the union of the founders of the Moscow Mathematical Society and its purposes in connection with the past of the universal and Russian consciousness [obshchechelovecheskogo i russkogo soznaniĭa]. <...> 23. The life and critique of pure consciousness; inspiration. <...> God-given and grace-filled first principles [blagodatnye nachala]. God-man testimony [bogochelovecheskoe svidetel'stvo] and its cultural and historical influence. Historians-prophets <...>—24. <...> Two kingdoms: the kingdom of right and law and the kingdom of falsehood.[5]

DOI: 10.1057/9781137338280

3

One of the few things easily discernible in the *Moscow Philosophic-Mathematical School* is the preoccupation of the author with what may be broadly called the idea of universal mathematics. Like Bugaev, however, Nekrasov avoids using these words. Nevertheless, he clearly believes that everything is constructed on mathematical principles ("Thou hast arranged all things by measure and number, and weight" is a favorite quotation with him[6]) and that mathematics is therefore universally applicable, both beliefs being attached to the tradition of *mathesis universalis.*[7]

Nekrasov was aware of this tradition,[8] but his main point of departure was Laplace, and in particular, the latter's *Introduction* to his *Théorie des probabilités* (1812). Theory of probability was described there as a science concerning the "most important questions of life" ("qui ne sont en effet, pour la plupart, que des problèmes des probabilités") and the entire system of human knowledge.[9] Laplace's treatment of theory of probability inspired Nekrasov's meditations on the universal character of mathematics.[10]

The *Moscow Philosophic-Mathematical School* provides another reason why the Moscow "school" has not been sufficiently researched. Constituting the most important source for its ideology, Nekrasov's work is at the same time such a quagmire of universality that the only escape is to celebrate the author's profoundness or to dismiss his absurdity. From the perspective of our research, it is, however, essential to give some impression of the philosophical and ideological content of this work.

Kantianism

1

Not far from the beginning of the *Moscow Philosophic-Mathematical School* we read that "the inner initiative of the cognizing spirit [vnutrenniaia initsiativa poznaiushchego dukha]"[11] is best to be seen in the operations of the geometer:

> <...> geometrical knowledge must take into account not only the observation but also the inner eye, judging metrically and autonomously [merno i avtonomno sudiashchim], <...> this inner measure of things must play a no less important legislative role in all the rest of the coordinations of the knowable world.[12]

DOI: 10.1057/9781137338280

The wording of this passage is clarified by the statement that geometry is a discipline which reveals in the most obvious way the nature of the "exact critical knowledge."[13] The word "critical" refers to Kantianism. Nekrasov's notion of it may have been very limited and inadequate,[14] but still it deserves attention. Even a mind of the most fantastic turn needs outside sources. Some of these, reflected in a characteristically distorted way in Nekrasov's writings, were part of contemporaneous intellectual trends. Kantianism was among those of prevailing interest at the time, and the philosophy of geometry was inseparably connected to it.[15]

Nekrasov elaborates on Kant's view of mathematics and geometry as the most graphic example of pure reason, drawing its material from itself without having recourse to experience. The "legislative role" assigned to the "inner eye" refers to Kant's famous assertion that "the understanding does not derive its laws from, but prescribes them, to nature." In Nekrasov's text, aiming to demonstrate the overwhelming importance of mathematics, it is not simply reason, but an inner sense of measure which organizes or, as he puts it, echoing Kantian vocabulary, "coordinates" the world. The operations of the geometer who proceeds according to the rules dictated by his "inner measure of things" are perceived as evidence in support of the Kantian theory of knowledge and are constantly associated with it.

2

There is a curious passage referring to this group of ideas and catching one's attention with its wild and poetic language:

> Every specific intellectual subject, morally orientating in the universe, builds for himself and his needs, according to his intellectual abilities, a notion of the diversity [mnogoobrazii][16], in which his own dazzling suns and stars shine for him. In this imaginary morally bipolar space everybody allots himself a particular corner, or rather a particular trajectory of his own personal existence, and similarly tuned subjects [odnorodno nastroennye sub"ekty], attracted to each other by a particular sympathy, founded on this moral consonance [moral'nom sozvuchii], create for themselves a path by a joint effort.[17]

Everybody is engaged in constructing imaginary spaces in the way geometers are, writes Nekrasov. It is thus that we create "moral universes," the qualities of which depend on their creators. Accordingly, everybody's "moral universe" is illumined by their "own dazzling suns

DOI: 10.1057/9781137338280

and stars" echoing Kant on the starry sky and the moral imperative from the *Critique of Practical Reason*. This passage is inserted in a section of Nekrasov's work where he is trying to propose a new version of the "theory of knowledge and the laws of reason," which he argues is connected with the notion of the "Absolute Being, His Reason or Wisdom [o Bezuslovnom Sushchestve, Ego Razume ili Premudrosti]."[18] Nekrasov's aim is to correlate the divine will with the Kantian creative "initiative" of those building their moral universes *more geometrico*.[19]

Mathematics forms the basis of our thinking, Nekrasov argues, whence the metaphor of everybody's moral life as the building of an imaginary geometrical space. The singularity of this mental picture finds its adequate expression in the language of the quoted passage. A student of *Petersburg* may indeed wonder whether this obsession with geometry, so colorfully rendered in a volume devoted to Belyĭ's father's memory, written by a mathematician–functionary, might be connected to our senator, who was in addition not unaffected by contemporaneous Kantianism.[20] Even if we cannot say with certainty that it was the extraordinary use of geometry in Nekrasov's writings that suggested to Belyĭ the geometrical motif of his novel, this has to be taken into account in any commentary to *Petersburg* aspiring to survey the most significant probable meanings implied by the author.

Platonism

1

It has been pointed out that Nekrasov was a "Platonist according to his philosophical views."[21] Now I shall concentrate on some of the features of his Platonism as displayed in the *Moscow Philosophic-Mathematical School*.

Nekrasov's Platonism was associated with his interest in Kant. Not only can we see this in the cited fragment on the "stars and suns" marked by the Platonic conception of the world harmony. The "inner measure of things" "coordinating" the world seems to have been linked in Nekrasov's mind to Plato. The former's use of "mernost'" (rhythm, regularity) and "mernyĭ" (measured, metrical), especially in such expressions as "metrical judgments [mernye suzhdeniĭa]," matches the Russian translation of Plato's *Republic* by Vasiliĭ Nikolaevich Karpov (1798–1867),[22] who employed these words to render the Greek

DOI: 10.1057/9781137338280

"ἐμμετρία" and "ἔμμετρος": "<...> do you think that truth is akin to measure [ἐμμετρίᾳ; mernosti] and proportion or to disproportion?"— "To proportion." [ἐμμετρίᾳ; mernosti]—"Then in addition to our other requirements we look for a mind endowed with measure [ἔμμετρον; rassudka po prirode mernogo] and grace whose native disposition will make it easily guided" (486D).[23]

The interweaving of Kant with Plato followed a bent of the nineteenth and early twentieth centuries to constantly compare the two, and indeed to frequently pronounce on their similarity. The omnipresence of Kantianism at the turn of the century made Plato especially relevant (the widespread taste for mysticism being another important factor, which was also not remote from the fashion for Kantianism).

For mathematicians interested both in philosophy and in the history of mathematics, as was common in Bugaev's circle, Plato was bound to be still more important for yet another reason. The view that Plato played a major role in the history of science, and in particular in mathematics as a methodologist, was then firmly established.[24] Even the tradition of considering Plato to have been an outstanding mathematician, which had already been refuted,[25] probably lived on in this circle. Finally, Plato's authority was particularly valuable for the Moscow mathematic–philosophical school, since he regarded mathematics as a basis for education and philosophy.

All this, reinforced by Nekrasov's conservative instinct to see his own ideas as belonging to an old and venerable tradition, turned Plato into an extremely relevant source. Nekrasov seems to have read him as a quite contemporary figure. This is not surprising, because Nekrasov was equally remote from what was going on around him, as he was from the ancient past. He misconceived Plato as contemporary because he misconceived the present. When the minister of education N. P. Bogolepov was assassinated in 1901, Nekrasov wrote an obituary using an epigraph from the *Apology of Socrates*,[26] drawing a parallel between the death of the minister and the celebrated death of the philosopher—not an easy allusion. Keeping in mind numerous passages of the *Moscow Philosophic-Mathematical School*, some of which will be cited below, this episode exemplifies Nekrasov's habit of thinking about state matters through the prism of Plato. This included representing the ruling power as a caste of philosophers, and not only in utopian projects, but through wishful (and clearly obsequious) thinking about the present.

DOI: 10.1057/9781137338280

2

Beyond all the complexities, we find in the *Moscow Philosophic-Mathematical School* the outlines of the Platonic state.

The state as a whole is compared by Nekrasov to a "judging individual [sudîashchaîa lichnost']" whose reason must control passions, which distinctly echoes the *Republic*. The resulting social harmony as described by Plato haunts the imagination of our author.[27]

Among the forces causing social disharmony, "sophists" play a prominent role. Interestingly, the term is not merely a conventional designation of demagogs employing deceptive arguments. In contemporary debates about the Russian–Japanese war and the criticism of the government Nekrasov hears the:

> *din* of the people of false principles [shum lîudeï lozhnykh nachal] advertising idols, and alluring agnostics in the nets of the false authorities, [which] always more or less drowns voices of the true popularizers of the exact knowledge and skill [vsegda bolee ili menee zaglushaet golosa istinnykh populîarizatorov tochnogo poznaniîa i umeniîa]. These gentlemen <...> *at all the crossroads of life (in the press, in parliaments, in corridors, in taverns, in carriages, in the streets, in dining-rooms etc.)* <...> *shout and scold* <...> [na vsekh perekrestkakh zhizni (v pechati, v parlamentakh, v korridorakh, v traktirakh, v vagonakh, na ulitsakh, v gostinykh i t.d.) <...> krichat i branîatsîa <...>]. Owing to *this vain din* [blagodarîa ètomu suetnomu shumu], in politics and journalism there have never governed <...> the terms of the exact knowledge <...>.[28]

Nekrasov's wording betrays the real source of the "din," that of the "multitude" manipulated by sophists in the *Republic*:

> <...> the multitude are seated together *in assemblies or in court-rooms or theaters or camps or any other public gathering of a crowd, and with loud uproar* censure some of the things that are said and done and approve others, *both in excess*, with *full throated clamor* and *clapping of hands*, and thereto the rocks and the region round about *re-echoing redouble the din* of the censure and the praise. (Plat. Rep. 6. 492bc; italics mine)

3

The way Nekrasov wrote of "Politics" seems to be suggestive of the same source: "Human politics is only one of the rays of the universal Politics, governed by the one world Sovereign Power and Its supreme right <...>."[29] "Universal Politics" here does not signify a course of political

DOI: 10.1057/9781137338280

life but the governing of the world, just as "πολιτεία" meant the govern-ing of the city; "human politics" is the governing of the state. The "one world Sovereign Power," spreading its "rays," is not in this case merely a Christian idea ("the Sun of Truth" is a frequent expression in Nekrasov), but also a reminder of the Platonic divine mind ruling the universe. In the same Platonic fashion Nekrasov draws a correspondence between the "one world Sovereign Power," which is divine and therefore entirely rational, and the sovereign of the state, described as a "bearer of the enlightened political logos [nositel' prosveshchennogo politicheskogo logosa]."[30]

As a reader of Plato, and some works on the history of Greek phi-losophy, Nekrasov knew that the first meaning of "logos" was that of computation and reckoning.[31] The sovereign in possession of "the enlightened political logos" suggests that just as in all the other spheres of life, governance of the state is fundamentally mathematical.[32]

4

Nekrasov's vision of the "truly rational state [istinno ratsional'noe gosudarstvo]," which he does not forget to distinguish from Plato's "abstract idealism"[33] (yet more evidence of Nekrasov's preoccupation with Plato) includes, as one of its most distinctive characteristics, the so-called "power of exact positive political knowledge."[34] It is defined as "an organ of cognizance of political things through pure political consciousness (political contemplation) [organ poznaniia politicheskikh veshcheĭ posredstvom chistogo politicheskogo soznaniia (politicheskogo sozertsaniia)]."[35] The more difficult it is to grasp what exactly is meant by "political contemplation," the more inevitably one is reminded of Platonic philosophers, whose capacity for contemplation trained with the help of mathematics marks them out as destined to be rulers.

5

Statistics is given an exceptionally important place in Nekrasov's vision of the "truly rational state." Everything in it, down to private secrets, must be registered and recorded in terms of the "secret symbols (ciphers) of <...> statistical vocabulary [pod skrytymi simvolami (shiframi) <...> statisticheskogo slovaria]."[36]

This may seem surprising. Statisticians had an often justified reputa-tion to be political radicals. Thus, after having visited Ukraine in spring

DOI: 10.1057/9781137338280

1902, Minister of the Interior V. K. von Plehve pointed out to Nicholas II that statisticians exercised a corrupting influence on the population. The question was then discussed in the press.[37] One of Nekrasov's points in his paper *The Philosophy and Logic of the Science of Mass Phenomena in Human Activity*, read at the sitting of the Moscow Mathematical Society in August of 1902, was the wrong-headed use of statistics by positivists.

This train of thought is continued in the *Moscow Philosophic-Mathematical School*, where statistics become part and parcel of Nekrasov's elaboration of the *Republic*. His "truly rational state" is to be covered by a network of statistical bureaus which must communicate their observations to "the central cognitive institutions [v tsentral'nye poznavatel'nye uchrezhdeniia gosudarstva] <…> giving the opportunity to the central statistician-epistemologist and the sovereign consciousness to draw the measure of things and hence the legislative and other motives."[38]

"The central cognitive institutions" supervised by "the central statistician-epistemologist," who must be both mathematician and philosopher, represent the "power of exact positive knowledge." "The measure of things" is taken from Platonic vocabulary, and reinforces the link with the body of ideas indicated above. Statistics for Nekrasov are not primarily associated with "facts." They are important as material for the work of the "Sovereign consciousness." The emphasis is placed here not on statistical data itself, but on statistical data as the means of functioning of a state that would be a closer equivalent to the "world Politics" bringing the "Sovereign consciousness" nearer to the ideal of the omniscient divine mind. Those statisticians who lay stress on "facts" have a share in Nekrasov's argument against dangerous troublemakers.[39]

Arithmology

Right in the opening pages of *Matematicheskiĭ sbornik* Nekrasov invokes the Creator and eternal essences.[40] Although interweaving such vocabulary into a work on mathematics might deter readers, the persistence with which these terms are used indicates the author's belief in harmony between the scientific pursuits of the Moscow mathematicians and Christian dogmas.

In this, Nekrasov did not merely follow his capricious fancy and eccentric tastes. Two factors which contributed to such tendencies are easy to grasp. The first is Bugaev's idea of arithmology.

DOI: 10.1057/9781137338280

1

In the oft cited work *Mathematics and the Scientific-Philosophical Conception of the World* ([*Matematika i nauchno-filosofskoe mirosozertsanie*], 1897[41]), in which Bugaev formulated his notion of arithmology, he makes the following remark, which can serve as an introduction to the theme of arithmology for the purpose of the present study:

> Les vérités de l'arithmologie portent en elles l'empreinte d'une individualité originale, vous attirent à elles par leur caractère mystérieux et leur beauté frappante. On n'explique que par ce fait pourquoi certains penseurs ont rattaché aux nombres entiers différentes questions de philosophie mystique.[42]

No kind of involvement in mysticism can be deduced from these words, which would rather point to the contrary. The very word "arithmology," which currently is closely associated with esoteric knowledge, was associated at that time with mathematics proper, rather than with "mystical mathematics."[43]

Nonetheless, Bugaev was aware of the latter, as the quotation indicates. His son mentions his "flight into Pythagoreanism [ulet v pifagoreĭstvo]," which most probably reflects not only Belyĭ's look at arithmology, but something he might have heard from his father.[44]

Contemporary mathematicians tended to disregard Bugaev's ideas on arithmology.[45] Nevertheless, as a notion designating a particular philosophical outlook it has gained some popularity and occurs in various historical works devoted to that period.[46] Moreover, despite obscurantist connotations which arithmology acquired in the writings of some of Bugaev's disciples, it reflected the contemporary drift towards the study of the discrete (exemplified by Ludwig Boltzmann's theories), which was to have a defining impact on the science of the following century.

2

Bugaev claimed that mathematical studies of the discrete functions (what he called arithmology in the narrow sense) heralded a new era of thought. The previous phase of scientific development was, we are told, defined by the domination of mathematical analysis. That was the only reliable groundwork for the whole body of learning. It taught the mind to seek for causes and not to think about purposes. As a result, determinism became the dominant scientific and philosophical outlook. Continuity, which the mind trained by mathematical analysis was quick

DOI: 10.1057/9781137338280

to see everywhere, came to be regarded as the necessary characteristic of all phenomena. Hence emerged Darwinism, Bugaev argued; the same habits of thought were transported to modern psychology and sociology. Bugaev believed the achievements of this scientific phase to be great and numerous, but that there was also a side effect of marked importance. "The analytical point of view" suggested that everything in the world is submitted to immutable laws ("à des lois analytiques, fixes et continues"); that nature knows neither good, nor evil, that "good and evil, beauty, justice and freedom" are only illusions of the human mind:

> Dans les considérations de certains philosophes est venue à prédominer le sentiment de la fatalité, de la nécessité absolue et inévitable. Le destin, le sort du monde antique se fait jour dans ces opinions.[47]

However, according to Bugaev, analysis, with its focus on continuous functions, forms only one part of mathematics. The other part is arithmology, which studies the discrete functions and thus is able to provide a firm basis for the study of the discrete. The latter is exemplified by the phenomena of individuality, by those of ethics and esthetics, all of which cannot be accounted for from the point of view of continuity and mathematical analysis. Having developed the tools for dealing with the discrete in mathematics, arithmology should become a firm scientific basis for a new worldview, which will readmit notions of good and freedom, argues Bugaev. Arithmology will dispel the current scientific fatalism.

Bugaev is rather vague about the concrete use of arithmology. He broadly envisages a complete change of the general scientific and philosophical outlook. The direction of the change is vaguely idealistic. Religion is never mentioned explicitly, and when "destiny" and "the fate of the ancient world" are evoked as a property of the contemporary scientific frame of mind that is to be conquered by the new "arithmological" promise of freedom, one is left to wonder whether the association with Christianity, naturally coming to mind, played any part in this reasoning.

It is, however, significant that Bugaev's project was interpreted as predominantly religious. Whereas mathematicians were skeptical about arithmology, in 1903 it found its way in the Christmas issue of the very important Russian conservative newspaper *Novoe vremîa* [*New Time*]. One of the most popular publicists, Mikhail Osipovich Men'shikov (1859–1918), hailed the idea of arithmology with great enthusiasm and perfect confidence in its being destined to support the failing Christian

DOI: 10.1057/9781137338280

faith. Readers of this issue, making their way through Christmas stories and discussions about the possibility to communicate with the other world, were also supposed to find comfort in Bugaev's theories:

> <...> in pure mathematics, events of extreme importance are under way... <...> Do you know what arithmology is? From the depth of a magnificent method elaborated with artistic subtlety, a new mathematics of an entirely different principle is developing, a new enormous branch [of mathematics], shaking no less than the law of causality—the granite continent of the modern worldview. Not every cause has a proportionate consequence, not everything is submitted to fate. There is a mystery ruling over the inevitable, and our individuality can be stronger than death. In the theory of numbers, the thought of scientists is probing the possibility of a human consciousness entirely different from the modern one, of an entirely different attitude towards the world, a possibility of the long lost, and deplored by many, faith in God, a faith not only traditional, not fostered by education, but faith in the sense of a philosophical certainty more incontestable than the law of causality <...> One branch of mathematics—analysis—has killed faith, but the other, the higher one—arithmology—leads to restoring the faith worthy of sages. The living Divine Being is to be proved not by the weak prattle of poets, not by the swearing of saints [ne slabym lepetom poètov, ne kliatvami sviatykh], but by the application to nature of a new mathematical method, as great as analysis has been. It is amazing, is it not [ne pravda li èto udivitel'no]?[48]

Later on Men'shikov returned to the same subject on a similar occasion. The next year, the Easter issue of *Novoe vremia* published his article *The Eternal Resurrection* [*Vechnoe voskresen'e*]. Here arithmology was interwoven with a still more impressive picture of the succession of world religions, starting with Buddhism (the joyless faith that poisoned the "most gifted of human races," the Aryans), and arriving at the scientific determinism that galvanized into life the ancient fatalism of the Greek and the Roman. The Christian message, which delivered the world from inexorable fate, was described in a very similar way to that of Bugaev's presentation of arithmology, and the latter was then evoked as the new and mighty support of the Christian doctrine.[49]

Such an interpretation of arithmology was hardly a complete misconception. Bugaev's careful words on the secrets of freedom incompatible with fatal determinism may have actually had Christian implications easily discernible to the contemporaneous ear. Peppering his discourse to Bugaev's memory with theological terms, part of Nekrasov's intention was perhaps to offer a hommage to the views of his teacher.

DOI: 10.1057/9781137338280

"Arithmology" and "arithmological" occur with great frequency in the *Moscow Philosophic-Mathematical School*. What they actually mean is often difficult to grasp, but they are tightly interwoven into the prevailing Orthodox message of the whole.[50]

"The divine order"

The second important factor behind Nekrasov's attempt to unite theology and mathematics was his preoccupation with theory of probability and statistics. If Bugaev's circle endowed mathematics with the widest possible frame of reference and attached to it profound philosophical meaning, including potentially strong religious implications, this was all the more so for scholarship in the field of probability theory. Such an atmosphere enabled one to be fully aware of, and to embrace, similar beliefs of one's predecessors.

1

One of Nekrasov's favorite references is Johann Peter Süssmilch (1707–1767), a Prussian pastor (as Nekrasov does not fail to remind us on several occasions) and an author of the famous *Göttliche Ordnung* (1741), of which it would be appropriate to cite the whole title, as it better conveys the idea of the work: *Göttliche Ordnung in den Veränderungen des menschlichen Geschlects, aus der Geburt, dem Tode und der Fortpflanzung desselben erwiesen*. Süssmilch regarded statistics as a means of pious contemplation of the works of divine providence. Süssmilch believed that statistics could reveal the working of divine law—for instance, the stable ratio between male and female newborns revealed by statistics contradicts the impression of birth and death being fortuitous and accidental.

By Nekrasov's time this attitude to statistics had become obsolete. Statistics had lost its religious connotations. Statisticians were supposed to collect "facts" and therefore were the very embodiment of positivism in its popular meaning. During a fancy dress ball in *Petersburg* we catch scraps of a conversation in which a professor of statistics takes part: "The annual consumption of salt by an average Dutchman... The annual consumption of salt by an average Spaniard..." What can be further from theology than this? One cannot even easily see any philosophical framework applicable to such problems—a framework which could be related to theology at least genealogically or typologically.

DOI: 10.1057/9781137338280

The professor's concentration on "facts" is mocked by pointing to their absurd minuteness.

Bearing in mind that statisticians were very important and prominent at the turn of the twentieth century, to the effect that the figure of the statistician is an image familiar to every student of that period (and despite Belyĭ's mockery this remarkably energetic figure encapsules one aspect of contemporaneous cultural and scientific flourishing), it is interesting that there could be a very different set of ideas associated with this field, whereby statistical data was not positivistic "facts" but a revelation of divine providence. One cannot help citing Nekrasov's words that statistics applied to history can "give the mind instructive <...> parables [pritchi] and omens [znameniĭa]" which, as the very choice of these words indicates, would guide our lives as the parables and proverbs [pritchi] of the Bible.[51] However incongruous this phrase may seem, it illustrates a certain position that was not exceptional in the history of statistics, and that to some extent we may grasp by correlating Nekrasov's obscure pronouncements with one of his sources, who was at the same time a prominent example for such a view. What is more, his other, and still more important source, shows that such views must have been better remembered by contemporary statisticians than it may seem now.

This source was Adolph Quetelet.[52] Unlike Süssmilch, who had been rather forgotten by that time,[53] only to regain popularity with historians later on in the twentieth century, Quetelet was still read.[54] His studies were as important for statistics in Russia as elsewhere. Nekrasov devoted a work to reexamining Quetelet *Filosofiĭa i logika nauki o massovykh proĭavleniĭakh chelovecheskoĭ deĭatel'nosti (peresmotr osnovaniĭ sotsial'noĭ fiziki Quetelet)* ([*The Philosophy and Logic of the Science of Mass Phenomena in Human Activity (a reexamination of the principles of social physics of Quetelet)*], 1902) whereby he embarked upon the new "ideological" phase of his career. It is here that he first proposed his version of religious theory of probability and statistics. We will have to return to this work later.[55] Here it is enough to note that in Quetelet we find passages close in spirit to the cited position of Süssmilch. Being common knowledge for historians of mathematics, they are much less known among historians of ideology:

> L'homme semble croire que la matière seule obéit à des principes immuables de mouvement et de conservation, comme si le Créateur avait laissé ses œuvres imparfaites et s'était moins occupé d'assurer la stabilité du monde moral que celle du monde physique.[56]

DOI: 10.1057/9781137338280

In fact, in the moral world

> nous trouvons au contraire une admirable harmonie qui, tout en laissant à l'homme sa libre faculté d'agir, l'a cependant limitée avec tant de sagesse, qu'elle ne peut entraver en rien les lois immuables qui président à la conservation des mondes comme à celle des plus simples éléments qui les composent. <…> je n'ai d'autre but que de montrer qu'il existe des lois divines, et des principes de conservation dans un monde où tant d'autres s'obstinent à ne trouver qu'un chaos désordonné.[57]

Whatever the difference between this clearly expressed argument and Nekrasov's confusing passages concerning religion, the connection of the latter to statistics and probability theory was not as archaic as it may seem. It rested on a layer of common knowledge.[58] This background makes his work in retrospect look less incompatible with (even if more provocative in) its particular historical context.

2

It will be remembered that Laplace used probability theory in a different spirit. In particular, he refuted the probability of miracles, including that of resurrection,[59] which explicitly challenged the Christian tradition. By a peculiar sleight of hand, or a still more peculiar blindness to the actual meaning of what he read, Nekrasov in the *Moscow Philosophic-Mathematical School* turned Laplace into his main authority on Christian mysticism.

This odd interpretation was prompted by the passage of the *Introduction* to *Théorie des probabilités*, in which Laplace, dealing with the problem of illusions in the estimation of probability, was brought to touch upon psychology. In particular, he wrote of the "sensorium ou siège de la pensée" which, upon becoming impressed by exterior objects, got modified in ways still unknown and mysterious. He compared the character of these modifications to the phenomena of light and electricity.[60]

It seems to have been Laplace's vocabulary which set Nekrasov's imagination to work. Nekrasov must have known of Newton's conception of space as the "sensorium Dei." In Nekrasov's earlier work "sensorium" emerges in connection with the Divine intellect or the Soul of the World sending its "rays" and "vibrations" throughout the world to be crossed with and perceived by "the rays of all the other intellects."[61] The fact that Laplace used the word "sensorium," which by Nekrasov's time was obsolete, was probably enough to suggest that Laplace advocated Newton's conception.[62]

DOI: 10.1057/9781137338280

No less important was Laplace's comparison of psychological processes with light and electricity. Those acquainted with writings on psychic phenomena could have read his account of the "invisible physiology," if detached from the context, as setting the familiar scene for spectacular actions of "currents," "rays," and "vibrations" in current descriptions of telepathy and the like.[63]

Whereas Laplace's aim in treating of psychology was to describe a psychological mechanism for the forming and persistence of superstitions, Nekrasov, influenced by contemporary psychical research, was trying to present *Introduction* to *Théorie des probabilités* as a source of his most fanciful deliberations.[64] Thus, warning the reader against the evolutionary approach to history, he recommended a withdrawal into the "cell of a hermit" as the best "laboratory" of historical knowledge. Far off from the everyday "noise," "sensorium," linked through "visible and invisible rays" with world consciousness, gives evidence of the truth of the Christian view of history.[65]

Autocracy

Unlike Plato's ideal state, the one to be found in the pages of *Matematicheskiĭ sbornik* is ruled by a sovereign. The "power of exact knowledge" is no more than "the 'eye' of the Monarch ['glaz' Gosudaria]," providing materials for his "political contemplation."[66] Platonic though it sounds, with its emphasis on contemplative and disinterested knowledge—Nekrasov is cautious enough to remark that those in charge of this power do not care for practical rulings, which would be greatly appreciated by Plato's philosophers indifferent to power—it is very far from Plato, the more so as Nekrasov's state is deeply Orthodox. The "'eye' of the Monarch" must "admit that grace is the natural source of life [priznaiushchiĭ blagodat' estestvennym istochnikom zhizni]."[67]

The very expression the "truly rational state" is characteristic and reveals all the difference between the original "fair city" in Plato and Nekrasov's version of it. The adverb "truly [istinno]" was in great vogue with Russian monarchists.[68] The idea of the "truly rational state" is clearly intended to appeal to the ruling Tsar, its Orthodoxy being a reflection of the Russian state system whereby the Tsar was the head of the church.[69]

DOI: 10.1057/9781137338280

1

From what has already been said, it is evident that the issue of the *Matematicheskiĭ sbornik* in question does not treat of mathematics as such, nor is it simply a treatise on the place of mathematics among sciences and in society. It comprises an obscure text of which one of the most distinctive features is ideological intensity.

This is the conclusion to which one comes after reading and rereading the *Moscow Philosophic-Mathematical School*. But when opening this work for the first time, it is with some astonishment that one reads its first paragraph:

> At the time when all Russia is excited by the events of the terrible war and our troops shed their precious blood defending honor and property of their motherland on the field of battle, at this time engulfed by the interests of war, the Moscow Mathematical Society had to call together admirers of Nikolaĭ Vasil'evich Bugaev deceased in May, in order to pay hommage to him as a hero of peace and peaceful labor.[70]

Right in the opening pages of *Matematicheskiĭ sbornik* we are given an unexpected reminder of the latest political developments. This is obviously meant to demonstrate the Russian patriotism and loyalty of the Moscow Mathematical Society. It could have been seen as a mere indication of the desire to strengthen the financial position of the society or as another manifestation of Nekrasov's remarkable sycophantic zeal. But the rest of the work, once we start better orientating ourselves in its bizarre language and notions, leads us to the conclusion that the mention of politics in the beginning of the work perfectly corresponds to its place in Nekrasov's thought.

2

The language of the cited passage is curiously familiar: "events of the terrible war [sobytiĭa groznoĭ voyny]," the troops shedding blood, "defending the honor and property of their motherland on the field of battle [zashchishchaĭa na pole brani chest' i dostoĭanie svoeĭ rodiny]," "hero of the peace and peaceful labor [geroĭ mira i mirnogo truda]"—all these expressions are now firmly associated with official Soviet language.[71]

From the point of view of the subsequent history of the Moscow Mathematical Society this resemblance is interesting. Nekrasov reminds us of the fact, which is both obvious and often neglected, that Soviet

DOI: 10.1057/9781137338280

vocabulary owed much to the previous political regime. The ideology of this regime, though to a much lesser degree than the Soviet one, also cared about what science should or should not do. The idea of making science a political matter was present there, and in the case of mathematics it is not among liberals or radicals, who are usually held responsible for the subsequent abuse of ideology in the Soviet times, that we find it, but among conservatives. In particular, the notion of the monarch as the representative of God was one of the reasons for trying to establish a close connection between political power and sciences. In Nekrasov the "carefully weighed word of the Tsar [Tsarskoe vzveshennoe slovo]" is linked to the "objective Consciousness, the Word embracing all the disciplines by measure and weight [ob"ektivnym Soznaniem, s Slovom, obnimaîushchim meroîu i vesom vse distsipliny]."[72]

Although Nekrasov was an exceptional case, in that he pushed the idea of the political significance of mathematics to its extreme, he was by no means alone, and we shall meet some further mathematicians with similar views later on in this book.

The Soviet accusations leveled against the Moscow mathematicians, mentioned initially, are not only instructive of the tragic absurdities of Soviet times. These accusations, vulgar and primitive as they were, probably had more complex roots than it would seem at first sight. To understand them fully, we must first review the history of the Moscow Mathematical Society. Some of its members, and at least one of its presidents, held essentially the same opinions on the political relevance of science as those that were preached during the Soviet times. These members were monarchists and were trying to use mathematics for the ends of defending the monarchy. More importantly, however, they thought mathematics politically relevant. Could the idea of the political usefulness of mathematics have had some impact on those who operated in the Soviet context and, although holding directly opposite political views, shared the same belief in their relevance to mathematics? Was this belief so clearly expressed in the course of the campaign against the "reactionary Moscow Mathematical Society" formed exclusively by the Soviet ideological framework, or was there some more specific motive related to the Moscow "school" and the ideological literature it had produced? Could one say that the very fact of the profound structural convergence of the ideas of the new Soviet ideologues and those of the Moscow "school" helped invite the campaign and the persecution? All these possibilities may be suggested here only as questions. Without

DOI: 10.1057/9781137338280

considering them it is, however, hardly possible to give an adequate account of the background of the tragic events which befell the Moscow Mathematical Society in Soviet times.

3

We feel ourselves in the presence of a Russian monarchist throughout the book. It is enough to look through the sources Nekrasov references to realize this. Some references, hardly to be expected in a mathematical journal, would be perfectly natural in a work on social or political matters written by a reactionary. Nekrasov evokes K. P. Pobedonostsev, the Ober-Procurator of the Holy Synod, who was generally regarded as the perfect incarnation of the conservative spirit of that period,[73] and cites a work by another well-known conservative and monarchist N. I. Chernîaev *Iz zapisnoĭ knizhki russkogo Monarkhista* [*From the Notebook of a Russian Monarchist*], published in an important monarchist journal *Mirnyĭ trud* [*Peaceful Labor*].[74] Other works printed in this journal are also cited.[75]

Moreover, there was a close link between Nekrasov's mathematical studies and his monarchism. We cannot dismiss the latter as a subject not related to the former and therefore comparatively superficial with regard to his ideas on mathematics. In fact, he thought of mathematics as a mighty weapon for supporting the monarchy, which was perhaps one of the reasons why he never tired of reminding readers of his conservative views, the latter being part of a mathematical argument for him. We shall examine a very singular train of thought whereby he believed he had discovered a mathematical proof for the rights of the Russian absolute monarchy. Before addressing this topic, one should briefly recall some traditional conservative implications of mathematics relevant for Nekrasov and the Moscow "school."

Traditional conservative implications of mathematics

Two features of mathematics would appear to have been particularly attractive for conservatives and have suggested to them the ideological potential of mathematics.

First of all, mathematics seems to deal with stable unchangeable laws. It draws, according to Plato, "the soul away from the world of

DOI: 10.1057/9781137338280

becoming to the world of being" (Rep. 7. 521d). The study of mathematics might therefore be presented as a valuable means of accustoming the mind to both divine and social laws. A properly trained mathematician is naturally immune to the thought of social change. "How could such a science, which is in itself like an eternal draft of the [divine] wisdom, lead to freethinking?",[76] exclaimed an obscure mathematician from Kazan University in his speech on the use of mathematics, delivered shortly after the victory in the Napoleonic wars and containing an exalted address to the "Most gracious Monarch," Alexander I.[77] The speech concluded (before the final invocation of the "most gracious sovereign") with a depiction of the universal choir praising the Almighty, and the entire discourse lauded mathematics as a miraculous means of achieving the frame of mind enthusiastically displayed by its author.

The second significant feature of mathematics, pointed to in the same discourse, is its ancient origins, "contemporary to the creation of the world," and the "immutability of its doctrine."[78] Or, as a sympathetic reviewer of Nekrasov put it almost a century after the Kazan speech, "the enviable reputation" of mathematics largely rests on its having remained untouched by the revolutions which have often ruined other sciences and made them start everything from the very beginning.[79]

Thus mathematics could be presented as both focusing on immutable truth, and itself being a paragon of constancy incompatible with revolutionary change.[80]

Arithmology as a mathematical foundation of autocratic rule and Leo Tolstoï's philosophy of history

To this general rhetorical framework, important for Nekrasov and other members of the Moscow "school" but not exclusive to them, came a more ingenious and original feature, related to Bugaev's arithmology, at the core of which lay the problem of continuity/discreteness.

Bugaev's enthusiastic belief that he had found the mathematical key to understanding the domain of discreteness was a defining factor behind his circle's ideological activity. Within the "arithmological" frame of reference, the discrete phenomena of justice and freedom, that is, social and political phenomena, were no longer regarded as beyond the scope of mathematics.[81]

DOI: 10.1057/9781137338280

It is not clear what specific influences Bugaev drew on here,[82] but unlike the "mathematical-ideological" effort of his circle, the focus on the continuity/discreteness problem was something they shared with their time, along with the wish to connect this problem to political issues. The controversy around Darwinism is an evident example.

We will address here another example of nineteenth century thinking about continuity, Tolstoĭ's philosophy of history, in order to elucidate the political implications of arithmology in the *Moscow Philosophic-Mathematical School*. Tolstoĭ's philosophy of history is relevant to the present study not merely due to the importance assigned to the problem of continuity in the philosophical part of *War and Peace*, but also because Moscow mathematicians were among those with whom Tolstoĭ discussed his reflections on historical and moral laws. In fact, Tolstoĭ turns out to be a valuable source for the reconstruction of some aspects of the ideological history of the Moscow "school."

1

In the section of *War and Peace* on history and mathematics, Tolstoĭ proceeds from a thesis of the absolute continuity of motion ungraspable for the human mind, which is bound to divide it into discontinuous elements. This division is arbitrary and at the root of most human errors, including those affecting the understanding of history: it causes historians to concentrate on the actions of arbitrarily selected individuals ("a king or a commander"), that is, "discontinuous elements" of history, instead of trying to grasp history in its continuity, the only way of discovering the "laws of historical movement."

That these laws can be found is evident, we are told, from the example of the branch of mathematics dealing with the infinitely small:

> This modern branch of mathematics, unknown to the ancients, when dealing with problems of motion admits the conception of the infinitely small, and so conforms to the chief condition of motion (absolute continuity) and thereby corrects the inevitable error which the human mind cannot avoid when dealing with separate elements of motion instead of examining continuous motion.
>
> In seeking the laws of historical movement just the same thing happens. The movement of humanity, arising as it does from innumerable arbitrary human wills, is continuous.
>
> To understand the laws of this continuous movement is the aim of history.
> <...>

DOI: 10.1057/9781137338280

Only by taking an infinitesimally small unit for observation (the differen-
tial of history, that is, the individual tendencies of men) and attaining to the
art of integrating them (that is, finding the sum of these infinitesimals) can
we hope to arrive at the laws of history.[83]

Various attempts have been made to interpret this passage. Its meaning is still elusive. It is very difficult, if at all possible, to imagine how
the "individual tendencies of men" could be ascertained and how to
attain the sought-for "art of integrating them."[84] The sources that could
have directed Tolstoĭ's thought (and therefore might help us explain
his meaning) have not been clearly established. Here one may only
note that the quoted passage seems to connect strongly to traditional
representations of infinitesimal calculus. That this passage can be
easily related to the history of calculus and its influence suggests that
Tolstoĭ might have drawn his inspiration from some popular accounts
of Leibniz.[85]

From the point of view of the present study, it is important that this
train of thought in *War and Peace* shows some affinities with a contemporary book by a member of the Moscow Mathematical Society. This book,
entitled *Obzor kampaniĭ 1812 i 1813 godov* ([*A Survey of the Campaigns of
1812 and 1813*], 1868), purported to reveal mathematical laws of wars and
belonged to Prince Sergeĭ Semenovich Urusov (1827–1897), the outstanding chess-player and a mathematician.[86] The latter half of the 1860s was a
period of intensive friendship between Tolstoĭ and Urusov, which must
have brought mathematics closer to Tolstoĭ's imagination.

That Urusov's ideas influenced Tolstoĭ while he pondered his philosophy of history is the first link between Tolstoĭ and the Moscow "school."
Nekrasov respectfully names Urusov amongst its founders.[87]

Prince Urusov was an interesting and colorful figure inclined to use
mathematics for peculiar ends (at one point, he was engaged in calculating "the law of the mortality rate of tsars [zakon smertnosti tsareĭ]"[88]),
and full of confidence in the significance of his calculations. Having "discovered" the laws of wars, he shared his enthusiasm with Tolstoĭ: "All of
history is so simple, that one cannot help wishing to write at once [chto
nevol'no khochetsîa napisat' teper'] what is going to happen tomorrow,
and after-tomorrow, etc."[89]

Entertaining such ideas, and probably not being a very good mathematician, the Prince's relationship with the Moscow Mathematical
Society, which at the end of the 1860s harbored neither a "philosophic-
mathematical school," nor, in all likelihood, any particular reverence

DOI: 10.1057/9781137338280

for titles, was complicated.[90] In his passionate search for mathematical solutions of the most unlikely problems, as well as in his Slavophilism, Urusov can be considered a predecessor of the Moscow "school."[91]

2

The next member of the Moscow Mathematical Society to be mentioned in connection with Tolstoï's reflections on continuity and historical laws is Bugaev.

One should observe that Bugaev's early views were similar. We do not know his opinions on the matter of historical laws, but there is some material in his archives which has a bearing on the subject.

First, in one of his lectures on theory of probability, Bugaev introduced the notion of infinitesimal calculus with the same reflections as Tolstoy on continuity and discreteness, the former being a universal feature of motion, the latter the result of our inability to grasp continuity:

> All natural changes happen continuously, time grows continuously [vremîa vozrastaet nepreryvno], and every phenomenon passing into another changes continuously. Because of a particular organization of our mind, however, we can neither imagine continuous figures, nor express them. In the transition of one phenomenon to another we do not see continuous change, but can only catch certain moments of it; likewise, taking two numbers, let it be 1 and 2, we shall never come to a continuous series, however many intermediate figures we place between them, the passage from 1 to 2 in our mind is always effectuated by leaps.[92]

Next, we know that the young Bugaev, a liberal addicted to reading *Otechestvennye zapiski*, was an admirer of P. L. Lavrov's *Mekhanicheskaîa teoriîa mira* [*A Mechanical Theory of the World*], published in this journal.[93] This work was a perfect manifestation of what Bugaev later characterized as the "analytical worldview."

Reading in Lavrov about "arbitrariness" driven away by immutable laws of modern science, which marked the end of magic,[94] one has—if not a direct corroboration of the supposition that Bugaev bought books on magic and similar subjects in a different spirit than that of a mere collector of curiosities—at least some further inkling, how Bugaev's arithmology of later times, which in a sense re-introduced "arbitrariness," could have influenced the attitudes of some of its adepts towards religion. Miracles postulated by the Christian tradition came to be regarded as scientifically grounded in arithmology.[95]

DOI: 10.1057/9781137338280

Finally, and most importantly, it was the problem of "laws" that Tolstoĭ discussed with Bugaev when they got acquainted. Tolstoĭ's diary offers us a glimpse into what seems to have been a transitional phase of Bugaev's thought, preceding his conception of arithmology. Here is the entry dated May 16, 1884:

> A delightful [prelestnaĭa] idea of Bugaev, that the moral law is the same as the physical one, but it is "im Werden." It is more than *im Werden*, it is recognized [on soznan]. Soon it will be impossible to put one into jail, to make wars, to glut oneself, fleecing the hungry [otnimaĭa u golodnykh] in the same way as now one cannot eat people, sell people.[96]

This entry is to be compared with Tolstoĭ's observations made three days later. One could imagine, he noted in a letter, that there was such a period when physical laws had not yet been as stable as they are now; it follows that one could look upon the moral laws as only temporarily lacking the immutability characterizing the physical ones.[97]

This suggests that the idea of laws had by that time become partly dissociated in Bugaev's mind from the idea of immutability. Though presumably stable in future, moral laws are compared with the once unstable natural ones. Thus, the law was, so to speak, set in motion. Thinking about laws was no longer grounded exclusively in terms of immutability,[98] which was probably a preliminary phase to shifting the focus from continuity (with its inevitable and fatal laws) to discreteness (with its "leaps" and freedom).

Anyway, there was a sharp contrast between Bugaev's early views embracing the absolute rule of natural laws and the universal character of continuity, and his late ideas centered on freedom and discreteness. This evolution was paralleled by the change of his political outlook. One may only wonder whether his early liberal position was in any way supported by his scientific beliefs, or whether his later monarchism was somehow linked to his arithmological speculations. What is certain is that, in the case of some of his disciples, these sets of ideas—on continuity and discreteness, on the one hand, and on politics, on the other—became intertwined.

3

Nekrasov took up Bugaev's emphasis on mathematics as a means of dealing with discreteness, or, as he liked to put it, with "freedom." The field in which he distinguished himself was theory of probability, which

DOI: 10.1057/9781137338280

he closely associated with arithmology, as another non-deterministic discipline. He made his debut as philosophizing mathematician with a work on mass phenomena, in which he tried to propose a mathematical proof of the freedom of will.

The Philosophy and Logic of the Science of Mass Phenomena in Human Activity was published in *Matematicheskiĭ sbornik* in 1902 while Bugaev was still president of the Moscow Mathematical Society, which probably tells us something about his own views. In this work Nekrasov asserted that probability theory, when applied to statistical data, furnished evidence of the freedom of will.[99]

On the one hand, this was a return to what Süssmilch had thought about the function of statistics, with additional references to the Russian Orthodox tradition. Nekrasov hoped to provide the Orthodox Church with a mathematical corroboration of one of its fundamental dogmas. Quite evidently, this was not only a religious issue, but also a political one, due to the position of the Russian Tsar within the hierarchy of the Orthodox Church.

Yet, the political implications of this work were deeper and more curious. Nekrasov does not seem to have worked some of them out before he wrote the *Moscow Philosophic-Mathematical School*. However, one of the sections of his study devoted to Tolstoĭ,[100] which now seems less surprising against the background of relationships between the latter and members of the Moscow Mathematical Society, outlined above, sheds light on a very particular train of thought we can trace in the *Moscow Philosophic-Mathematical School*.

Nekrasov starts this section by criticizing Tolstoĭ's "belittling of the significance of personal influences of the main characters" of *War and Peace*: "These personal influences *are effaced* here <...>, and the role of above-average individuality loses value [Èti lichnye vliĭaniĭa *stushevyvaĭutsĭa* zdes', <...> i lichnost' vyshe srednego urovnĭa obestsenivaetsĭa v ee roli]."[101]

Tolstoĭ memorably thus phrased one of his conclusions, drawn from the postulate of the "absolute continuity of motion" and linked to his thesis that the "sum" of the individual strivings of men is the driving force of history:

A king is history's slave.

History, that is, the unconscious, general, swarm-life of mankind, uses every moment of the life of kings as a tool for its own purposes.[102]

DOI: 10.1057/9781137338280

Or, as Tolstoĭ goes on to say about Napoleon, the latter was "in the grip of inevitable laws."

The stress on continuity connects with the emphasis on laws leaving no room for freedom of will. Those who exemplify this total submission to "inevitable laws" most clearly are precisely those traditionally regarded as the main movers of history, namely, military leaders and kings.

Reading the quoted passages from the *War and Peace*, one can easily see why continuity was potentially irritating for a monarchist. Within its framework "kings" were transformed into "slaves" obedient to laws, even if only the laws of history.

On the other hand, the stress on discreteness provided a precious breach in the intractable rule of laws. According to arithmology, nature is full of creative "leaps" not explicable by laws of causality. The "divine order" presupposes freedom of will, which, Nekrasov believed, he had proved in his cited work on mass phenomena. But he did not stop there. By the time Nekrasov published the *Moscow Philosophic-Mathematical School* he had "noticed"—and Tolstoĭ had perhaps prompted his thought in this direction—that his mathematical proof of freedom of will could be used as an argument for absolute monarchy. If free will exists, and has at last found its ingenious mathematical confirmation, the same magical formula restores the autocracy of the monarch (reduced to an illusion in the framework of continuity) to the status of a God-sanctioned reality. Autocracy is the highest manifestation of the natural (God-given) freedom of will.

This is why the very notion of free will comes to be associated for Nekrasov with sovereignty.[103] Even writing on matters remote from things political, such as psychology, he describes the will of an individual as "an enlightened beneficial will [prosveshchennaĭa blagaĭa volĭa],"[104] employing the contemporaneous terms of the monarchist vocabulary.

4

This explains, at least partly, some of the most obscure passages in the *Moscow Philosophic-Mathematical School*. Here is one of them, translating the Platonic harmonious state into the context of the Russian Autocracy:

> The historically developed sovereign power of the Russian state, built upon the Russian Orthodox live ideal [na pravoslavno-russkom zhivom ideale], is the soul of the harmony of the divided powers, uniting them and

DOI: 10.1057/9781137338280

conducting the ethic-social chorus of the state. It is the harmony of this chorus which is the live balance of the divided powers [zhivoe ravnovesie razdelennykh vlasteĭ], moderating independence of isolated parts. Blind independence and blind consolidation are here replaced by the mind-measured and law-measured freedom, necessary for every live element of the wisely conducted chorus and subjected to the live and the most *autonomous autonomy*, the mighty Sovereign-Autocrat [zhivoĭ avtonomneĭsheĭ avtonomii, moguchemu Gosudarĭu Samoderzhtsu].[105]

The presumed "harmony of the divided powers" referred to the burning issue of the Russian political life, namely the absence of any united government. There was no coordination of the actions of different state departments. Nicholas II, following the example of his father, feared the consolidation ("the blind consolidation," in Nekrasov's terms) of ministers and preferred to receive their reports in private, which led to the Council of Ministers being a weak and ineffective organ.[106]

Calling the Tsar the "most autonomous autonomy," Nekrasov followed Bugaev's suggestion (prompted by Kantian notion of "moral autonomy") to replace the question of freedom of will with that of autonomy of individuals. Unlike "freedom," "autonomy" was impossible to confuse with "lawlessness" or "arbitrariness [proizvol]."[107] It is against this background that Nekrasov's monarchism is to be clarified:

> The Autonomy of the Monarch, the sovereign bearer of the enlightened political logos, is a shrine, the national sacred arc of the covenant before which all the truly beneficial autonomies of the state reverently bend their heads [Avtonomiia Gosudaria, suverennogo nositelia prosveshchennogo politicheskogo logosa, iavliaetsia sviatyneiu, gosudarstvennym sviashchennym kivotom zaveta, pered kotorym blagogoveino skloniaiut svoi golovy vse istinno blagotvornye avtonomii gosudarstva].[108]

There was a slightly extenuating circumstance behind Nekrasov's including such passages, an odd combination of philosophical notions with official clichés of that time which never fail to convey an impression of self-parody, in a book on the mathematical "school." He evidently believed he had found a way of defending the absolute monarchy on mathematical grounds. It was indeed a mathematical matter with him, hinging on his mathematical proof of the freedom of will and intimately connected with the whole philosophy and mathematics of Bugaev's "school," based on the latter's arithmology. However puzzling Nekrasov's emphasis on his monarchism in the *Moscow Philosophic-Mathematical School* may seem, it was not without some kind of logic.

DOI: 10.1057/9781137338280

This logic is, in great part, that of juggling words, whereby the will of the emperor equals free will, and the laws of nature suggest the laws of the state. Even Nekrasov's preoccupation with epistemology was imbued with his loyal feelings. Whereas the Tsar is the "most autonomous autonomy," it is the "inner eye" "judging autonomously" which must play a "legislative role in all the coordinations of the knowable world."[109] The "inner eye," lending its laws to nature, is "autonomous": it is not subordinate to laws, and thus represents an epistemological counterpart, that Nekrasov regards as a corroboration, of a view of the Monarch who (with the help of the "power of exact knowledge," that is, the "'eye' of the Monarch"[110]) lays down laws without being subordinate to them.[111] Once again, mathematics as the most graphic manifestation of this epistemological principle turns out to be the theoretical buttress of monarchy.

5

The main thrust of the *Moscow Philosophic-Mathematical School* appears to be the idea that mathematics pertains to (and is indeed essential for) political ideology. All the other themes of the work are dominated by this central argument. Nekrasov is trying to formulate a universal doctrine firmly based on mathematics and embracing the whole body of knowledge, including philosophy, theology and social science. This new doctrine has immediate relevance for contemporaneous Russian politics and the debates about monarchy at its core. Mathematics supports the monarchy not only indirectly, through its support of the Orthodox teaching, but in the most immediate way, by providing a mathematical foundation for autocracy, a foundation no doubt illusory, but in a sense logical, according to the peculiar logic that guides the succession of images in a dream. Discovering the nature of this logic adds to our knowledge about the mental and emotional atmosphere of the "school."

Notes

1   According to Belyĭ, Bugaev was not particularly fond of his pupil. See Andreĭ Belyĭ, *Na rubezhe dvukh stoletiĭ* (Moskva: Khudozhestvennaîa literatura, 1989), 73; id., *Moskva* (Moskva: Sovetskaîa Rossiîa, 1989), 36–37; Nekrasov is there called "Blagolepov".

2   Among numerous articles written on Nekrasov, the following were particularly useful to me: Eugene Seneta, "Statistical Regularity and Free

Will: L.A.J. Quetelet and P.A. Nekrasov," *International Statistical Review* 71: 2 (2003), 319–334; Oscar B. Sheynin, "Nekrasov's Work on Probability: The Background," *Archive for History of Exact Sciences* 57 (2003), 337 – 353.

3   Sheynin, "Nekrasov's Work on Probability," 338.

4   On Nekrasov's language, see Sheynin, "Nekrasov's Work on Probability," 340–342, and the references there given.

5   P.A. Nekrasov, "Moskovskaîa filosofsko-matematicheskaîa shkola i ee osnovateli," *MS* 25: 1 (1904), 3–4.

6   Non-canonical book Wisdom of Solomon, 11.21. In the *Moscow Philosophic-Mathematical School* it is to be found right in the beginning (6).

7   Written later than Nekrasov's work, E.V. Spektorskiĭ's *Problema sotsial'noĭ fiziki v XVII stoletii* ([The problem of social physics in the seventeenth century], 1910–1917) allows one to see what somebody interested in the tradition of *mathesis universalis*, could have known about it at that time. Spectorskiĭ's study started the history of sociology not from Comte, but from the then largely forgotten thinkers of the seventeenth century whose sociological theories were integrated into the framework of *mathesis universalis*. Spectorskiĭ's view of them as more profound and scientific than current positivism coincided with that of the Moscow "school." His survey of seventeenth century thought included interesting and sympathetic pages on theology. It was perhaps not an accident that Spektorskiĭ, who was one of the favorite pupils of A.L. Blok (the notable professor of law and the father of the poet) in Warsaw, defended the first volume of this work as his MA thesis in 1911 at Ĭur'ev (Dorpat) University: rector of the University at that time was the member of the Moscow "school" V.G. Alekseev (more on him see Chapter 4, Section "V.G. Alekseev"). On Spektorskiĭ, see A.A. Ermichev, "E.V. Spektorskiĭ (1875–1951). Biobibliographicheskaîa spravka," in E. Spektorskiĭ's *Problema sotsial'noĭ fiziki v XVII stoletii*, T. II (Sankt-Peterburg: Nauka, 2006), 506–523.

8   It follows, in particular, from his remark that the foundations of the social sciences were laid not by Auguste Comte, but by the philosophers and mathematicians of the seventeenth–eighteenth centuries. See "Otvet P.A. Nekrasova na zamechaniîa i vozrazheniîa otnositel'no filosofskikh i logicheskikh osnovaniĭ sotsial'noĭ fiziki," *VFiP* 68: 3 (1903), 583–584.

9   "On peut <...> dire, à parler en rigueur, que presque toutes nos connaissances ne sont que probables; et dans le petit nombre des choses que nous pouvons savoir avec certitude, dans les sciences mathématiques elles-même, les principaux moyens de parvenir à la vérité, l'induction et l'analogie, se fondent sur les probabilités, en sorte que le système entier des connaissances humaines se rattache à la théorie exposée dans cet essai" (Laplace, *Oeuvres complètes*, T. 7 (Paris: Gauthier-Villars, Imprimeur-libraire, 1886), V).

DOI: 10.1057/9781137338280

10   For an explicit reference to Laplace as an example of philosophical and mathematical synthesis, see "Otvet P.A. Nekrasova," 594.

11   Nekrasov, "Moskovskaîa filosofsko-matematicheskaîa shkola," 12.

12   ibid., 13.

13   ibid., 12.

14   Thus, evoking "theory of knowledge" *ad nauseam*, he understands it—at least in most cases—not as epistemology, but as a set of presuppositions guiding one's consideration of a subject.

15   Nekrasov was the member of the Moscow Psychological Society. The journal of the society *Voprosy filosofii i psikhologii* regularly published materials related to Kantianism.

      Nekrasov's sources on the philosophy of geometry included G.I. Chelpanov's *Predstavlenie prostranstva s tochki zreniîa gnoseologii* ([*The Notion of Space from the Point of View of Epistemology*], 1904; Nekrasov, "Moskovskaîa filosofsko-matematicheskaîa shkola," 13 n.)

16   "Mnogoobrazie" is one of Nekrasov's favorite words helping to "translate" non-mathematical subjects into the language of geometry. Being a part of everyday language, it is also employed both in philosophy (in this passage Nekrasov has in mind Kant's use of it), and as a mathematical term.

17   Nekrasov, "Moskovskaîa filosofsko-matematicheskaîa shkola," 46–47.

18   ibid., 43.

19   Despite Nekrasov's evident debt to Spinoza, the latter is characteristically treated as shallow and primitive (Nekrasov, "Moskovskaîa filosofsko-matematicheskaîa shkola," 22–23).

20   The senator is an "inborn" Kantian: he is said to confuse Comte with Kant, but his thoughts are often reminiscent of Kantian meditations of his son. Moreover, thinking about time, that is about himself (see Chapter 1, n. 23), Ableukhov the elder reproduces Kant's observations in the *Critique of Pure Reason*: "wir <...> stellen die Zeitfolge durch eine ins unendliche fortgehende Linie vor" (Der tranzendentalen Ästhetik, § 6 b); cf. the comparison of time to the endless "rectilineal prospect" in geometrical meditations of the senator (see Appendix 1).

21   M.V. Chirikov and O.B. Sheînin, "Perepiska P.A. Nekrasova i K.A. Andreeva," *IMI* 35 (1994), 124; http://www.sheynin.de/download/2_Russian%20 Papers%20History.pdf, 70.

22   Insisting that the ideal of the Moscow Mathematical Society was not that of the mathematician—"reckoner [schetchik]" but an "educated philosopher," Bugaev also kept in mind Plato's *Republic* in Karpov's translation; the latter used "schetchik" as an equivalent of λογιστικός (Plat. Rep. 7, 525 b; *Sochineniîa Platona*, trans. [V.N.] Karpov, Ch. III, Politika ili Gosudarstvo (Sankt-Peterburg, 1863), 370). See *Rech' prezidenta Moskovskogo Matematicheskogo Obshchestva N.V. Bugaeva* (Moskva, [1896]), 6; "Zasedanie

DOI: 10.1057/9781137338280

Moskovskogo matematicheskogo obshchestva," *MS* 21: 3
(1900), 538.

23 I have used the translation of the *Republic* by Paul Shorey (1934) to be found
in Perseus Digital Library: http://www.perseus.tufts.edu/hopper/. (The
electronic version is based on *Plato In Twelve Volumes*, Vols. 5–6 (Cambridge,
MA – London: Cambridge University Press – William Heinemann Ltd.,
1969).
Nekrasov's notion of "metrical" elements of thought combined Plato (hence
the vocabulary suggested by Karpov's translation), Herbartian mathematical
psychology (see Chapter 1, Section "Psychology"), and Laplace's observations
concerning probabilistic "instinct" (see Chapter 3, Introduction).

24 See L. Zhmud, *The Origin of the History of Science in Classical Antiquity,* trans.
Alexander Chernoglazov (Berlin – New York: Walter de Gruyter, 2006),
82–83.

25 ibid., 82.

26 P.A. Nekrasov, *Nikolaï Pavlovich Bogolepov* (Moskva, 1901).

27 Nekrasov, "Moskovskaîa filosofsko-matematicheskaîa shkola," 29, 81,
84, 97–98, 99, 106, 120, 121, 131–132 etc.; cf. [V.N. Karpov], "Vvedenie," in
Sochineniîa Platona, 19.

28 Nekrasov, "Moskovskaîa filosofsko-matematicheskaîa shkola," 91 (italics
mine).

29 ibid., 106.

30 ibid., 157.

31 For the mention of S.N. Trubetskoï's *Uchenie o Logose v ego istorii* (1900), see
Nekrasov, "Moskovskaîa filosofsko-matematicheskaîa shkola," 118 n.2. The
persistent use of "logos" in the *Moscow Philosophic-Mathematical School* as
a synonym of thought serves as yet another reminder of the mathematical
basis of mental life.

32 Following Plato, as he did, Nekrasov did not forget dialectics, which kept
regularly emerging in the *Moscow Philosophic-Mathematical School*. Unlike
Plato, however, it is mathematics that Nekrasov considered to be the
dominant science, and not dialectics, which he usually evoked as a synonym
of a pseudo-science.

33 Nekrasov, "Moskovskaîa filosofsko-matematicheskaîa shkola," 169.

34 ibid., 97.

35 ibid., 154.

36 ibid., 177. Statisticians would be entitled to learn private secrets, not for the
sake of repression, as it is pointed out, but like priests or doctors who do not
divulge information they receive.

37 See *Krizis samoderzhaviîa v Rossii 1895–1917* (Leningrad: Nauka,
Leningradskoe otdelenie, 1984), 122.

38 Nekrasov, "Moskovskaîa filosofsko-matematicheskaîa shkola," 176.

DOI: 10.1057/9781137338280

39   ibid., 156. In the notorious collected articles *Na bor'bu za materialisticheskuĭu dialektiku v matematike* ([*To Arms for the Materialistic Dialectics in Mathematics*]; Moskva-Leningrad: Gosudarstvennoe nauchno-tekhnicheskoe izdatel'stvo, 1931), which was one of the steps in the Soviet campaign against the Moscow Mathematical Society, there was published an article "Platon kak matematik" by a well-known mathematician and historian of mathematics Mark ĬAkovlevich Vygodskiĭ (1898–1965). This article, devoted to a refutation of the legend of Plato as a great mathematician, was an organic part of the campaign against the Moscow philosophic-mathematical school. It is in this context that Vygodskiĭ's presentation of Platonism as a reactionary philosophy of the exploiter classes should be regarded. On the last pages the author evoked contemporary Platonists, Nekrasov being one of them (181–182).

40   P.A. Nekrasov, "Predislovie," *MS* 25: 1 (1904), XII.

41   It was originally a paper read at the mathematical congress in Zürich, in French, in 1897. See N. Bougaïev, "Les mathématiques et la conception du monde au point de vue de la philosophie scientifique," in *Verhandlungen des ersten internationalen Mathematiker-Kongresses in Zürich vom 9 bis 11 August 1897* (Leipzig: Teubner, 1898), 206–223. The following year it was read at the Psychological Society in Moscow. See N.V. Bugaev, *Matematika i nauchno-filosofskoe mirosozertsanie* (Moskva, 1899).

42   Bougaïev, "Les mathématiques et la conception du monde," 209.

43   It seems to have been introduced into the Russian scientific language via French: "Arithmology <…>. This is the name which Ampère gave to pure mathematics in his 'Essai sur la philosophie des sciences' <…>" (F.A. Brokhaus and I.A. Efron, eds, *Ėntsiklopedicheskiĭ slovar'*, T. II (S.-Petersburg, 1890), 100). This definition echoed French dictionaries which referred to Ampère's classification of sciences proposed in the cited essay, as the source of the term "arithmology": "Arithmologie <…> (mot créé par Ampère). Science générale des nombres et de la mesure des grandeurs quelles qu'elles soient" (Paul Guérin, ed., *Supplément illustré du Dictionnaire des dictionnaires. Lettres, sciences, arts, encyclopédie universelle*, T. 7 (Paris: Libraires-imprimeries réunies, 1895), 91).

Bugaev's use of "arithmology" was different from that in Ampère who divided mathematics into "arithmologie" and geometry, the former embracing, among other branches of mathematics, mathematical analysis which Bugaev opposed to "arithmology." See André-Marie Ampère, *Essai sur la philosophie des sciences, ou exposition analytique d'une classification naturelle de toutes les connaissances humaines* (Paris: Chez Bachelier, Imprimeur-Libraire pour les Sciences, 1834), 40–42, 274.

44   Andreĭ Belyĭ, *Na rubezhe dvukh stoletiĭ* (Moskva: Khudozhestvennaĭa literatura, 1989), 61. Cf.: "The new theory of numbers is a return to

DOI: 10.1057/9781137338280

Pythagoreanism; and my father knew this" (Andreĭ Belyĭ and Aleksandr Blok, *Perepiska, 1903–1919* (Moskva: Progress-Pleĭada, 2001), 435).

45 As V. Kagan writes in his encyclopedia article, "the best Russian mathematicians—P.L. Chebyshev, N.A. Korkin and A.A. Markov—were not inclined toward these [Bugaev's] metaphysical constructions, argued their groundlessness, and did not even recognize any 'arithmology' [ne byli sklonny k ètim metafizicheskim postroeniam, dokazyvali ikh nesostoĭatel'nost', ne priznavali dazhe nikakoĭ 'aritmologii']" ("Bugaev Nikolaĭ Vasil'evich," In *Bol'shaia Sovetskaia Èntsiklopediia*. T. 7 (Moskva: Gosudarstvennoe slovarno-èntsiklopedicheskoe izd. "Sovetskaia Èntsiklopediia," OGIZ RSFSR, 1927), 770).

46 The fullest account of Bugaev's "arithmology" is a contemporary one, written by one of the members of the Moscow "school": W.G. Alexejeff, "Über die Entwickelung des Begriffes der höheren arithmologischen Gezetsmässigkeit in Natur- und Geisteswissenschaften," *Vierteljahrsschrift für wissenschaftliche Philosophie u. Soziologie* 3: 1 (1904), 73–92.

47 Bougaïev, "Les mathématiques et la conception du monde," 214.

48 M. Men'shikov, "Zvezdy i chisla," *Novoe vremia*, № 9990, December 25 (January 7), 1903, 7. The article was republished in full in ÎU. M. Koliagin and O.A. Savvina, *Matematiki-pedagogi Rossii. Zabytye imena. Kn. 4. Nikolaĭ Vasil'evich Bugaev* (Elets: EGU im. I.A. Bunina, 2009), 100–102.

49 M. Men'shikov, "Pis'ma k blizhnim. Vechnoe voskresen'e," *Novoe vremia*, № 10081, March 28 (April 10), 1904, 3–4.

50 P.A. Florenskiĭ suggested that the contemporary intellectual trends dealing with discontinuity and thus destroying the modern scientific outlook, arithmology being one of them, foreshadow the proximity of Apocalypse. See his *Stolp i utverzhdenie istiny*, T. 1 (I) (Moskva: Pravda, 1990), 127. A train of thought behind such expectations could be the following: a scientific grasp of phenomena escaping immutable physical laws might usher in a period when these very laws will cease to exist. See also Avril Pyman, *Pavel Florensky: A Quiet Genius. The Tragic and Extraordinary Life of Russia's Unkown Da Vinci* (New York – London: Continuum, 2010), 77, 277 n. 41. According to S.M. Solov'ev, Bugaev "plunged into the theory of numbers" would "fall into a sort of mystical frenzy" over the Apocalypse. See S. Solov'ev, *Vospominaniia* (Moskva: Novoe literaturnoe obozrenie, 2003), 170.

51 Nekrasov, "Moskovskaia filosofsko-matematicheskaia shkola," 79.

52 See Seneta, "Statistical Regularity and Free Will."

53 It was not, however, a complete oblivion. Süssmilch was not absent from books on the history of statistics. At the time of Bugaev's youth readers of *Otechestvennye zapiski*, which regularly printed materials on statistics, could find a sympathetic account of Süssmilch's ideas: A. Korsak, "Statisticheskie

DOI: 10.1057/9781137338280

vyvody ob obshchikh zakonakh narodonaseleniia i ego zhizni," *OZ* 139 (1861), 2–3.

54   Writing on Nekrasov, the outstanding statistician Ladislaus von Bortkiewicz characterizes Queteletism as "in a sense <...> obsolete" and having "no part in current Russian thought." See P.A. Nekrasov, *Theory of Probability* compiled, translated and commented by Oscar Sheynin: http://www. sheynin.de/download/5_Nekrasov.pdf, 62. The qualification "in a sense" seems significant. Quetelet's works were still widely known, and there was even some demand for translations. In 1911–1913 in Kiev there appeared the translation of *Physique social*.

55   See Section "Arithmology as a mathematical foundation of autocratic rule and Leo Tolstoi's philosophy of history."

56   Ad. Quételet, *Du Système social et des lois qui le régissent* (Paris: Guillaumin et Cie, Libraires, 1848), 103.

57   ibid., 9; cf.: " <...> rien n' échappe aux lois imposées par la toute-puissance divine aux êtres organisés" (Quételet, *Du Système social et des lois qui le régissent*, 16). See also Seneta, "Statistical Regularity and Free will," 323.

58   Another quite frequent reference in Nekrasov is the theologian and statistician Alexander von Oettingen (1827–1905). See also Chapter 1, n. 15.

59   Laplace, *Oeuvres complètes*, LXXXII–LXXXIII. Cf. M.Î.A. Vygodskiĭ, "Matematika i ee deiateli v Moskovskom universitete vo vtoroĭ polovine XIX v.," *IMI* 1 (1948)," 180–181.

60   Laplace, *Oeuvres complètes*, CXXIII–CXXIV.

61   P.A. Nekrasov, "Logika mudrykh liudeĭ i moral'. (Otvet V.A. Gol'tsevu)," *VFiP* 70: 5 (1903), 926. This work was in many ways a preliminary to the *Moscow Philosophic-Mathematical School*.

62   Nekrasov referred to Laplace on an earlier mention of "sensorium" in the same work (907).

63   One may refer to one of Bugaev's favorite authors, Camille Flammarion (1842–1925; by the moment of drawing up the catalog of his library he had three books by Flammarion: *Les mondes imaginaires et les mondes réels*, ed. 1868; *Contemplations scientifiques*, ed. 1870; *Vie de Copernic et histoire de la découverte du système du monde*, ed. 1872; ORK i R NB MGU, f. 41, op.1, ed. khr. 252, L. 78 ob., 8). His later work *L'inconnu et les problèmes psychiques* (1900) must have been known in Bugaev's circle. Like most of Flammarion's books it was translated into Russian; see a very favorable review of the book in *Bogoslovskiĭ vestnik*, which was also favorable to Nekrasov's ideas: Sergeĭ Kuliukin, "Îavleniia telepatii i znachenie ikh v oblasti osnovnykh psikhologicheskikh voprosov," *Bogoslovskiĭ vestnik* 2: 5 (1901), 200–227. The chapter entitled "L'action psychiques d'un esprit sur un autre" presents the same notion, widespread at that time, of how the knowledge of things distant

DOI: 10.1057/9781137338280

or past could be obtained, massively using the comparisons of rays, currents, inductions etc.

64　Nekrasov quotes Laplace's discussion of the invisible "physiology" in full ("Moskovskaîa filosofsko-matematicheskaîa shkola," 26–27 n.)

65　Nekrasov, "Moskovskaîa filosofsko-matematicheskaîa shkola," 69.

66　ibid., 154.

67　ibid.

68　"Istinno" seems to have become especially frequent during the reign of Nicholas II, who was addicted to this word, and this, according to S.ÎU. Witte, was at the root of its popularity at that time. See S.ÎU. Witte, *Vospominaniîa*, T. 2 (Moskva: Izd. sotsial'no-èkonomicheskoǐ literatury, 1960), 306, 501.

69　Here perhaps lies the explanation why Nekrasov is so critical of Plato whom he evokes on a number of instances in the *Moscow Philosophic-Mathematical School* almost exclusively to point out his deplorably "abstract idealism." A monarchist as he was, influenced by Platonism and probably well aware of this, Nekrasov may have been avoiding not so much the possible impression of dependency upon a particular source (which was well-known, and therefore immediately recognizable) as that of espousing a suspicious political doctrine.

70　Nekrasov, "Moskovskaîa filosofsko-matematicheskaîa shkola," 4–5.

71　The resemblance of Nekrasov's language and ideology to those of the Soviet times was pointed out in M.A. Prasolov, "'Tsifra poluchaet osobuîu silu' (Sotsial'naîa utopiîa Moskovskoǐ filosofsko-matematicheskoǐ shkoly)," *Zhurnal sotsiologii i sotsial'noǐ antropologii* 10: 1 (2007), 46–47.

72　Nekrasov, "Moskovskaîa filosofsko-matematicheskaîa shkola," 158.

73　ibid., 94 n.

74　ibid., 158 n.

75　ibid., 118 n. 7, 179–183.

76　[G.B.] Nikol'skiǐ, *Slovo o pol'ze matematiki* ([Kazan', 1816]), 42. I found this tiny brochure following a reference in Alexandre Koyré's early book *La philosophie et le problème national en Russie au début du XIXe siècle* (Paris: Librairie ancienne Honoré Champion, 1929), 74–75. Let us note that Nikol'skiǐ had the opportunity to realize the notions which he formulated. Under the notorious Mikhail Leont'evich Magnitskiǐ (1778–1844) charged with functions of warden of Kazan School District, he became rector of Kazan University. It will be remembered that Magnitskiǐ almost succeeded in destroying this university by trying to turn all the disciplines into religious and loyal ones (Koyré, *La philosophie et le problème*, 70–76). This attempt somewhat resembles what the Moscow "school" would try to do, the more so since the head of Kazan University was a mathematician embracing the same beliefs in the ideological utility of his subject. Nekrasov almost certainly knew about

DOI: 10.1057/9781137338280

Nikol'skiĭ; he could hardly have missed a brief account of his ideas in *VFiP*. See Al.V. Vvedenskiĭ, "Sud'by filosofii v Rossii," 42: 2 (1898), 331.

77 Nikol'skiĭ, *Slovo o pol'ze matematiki*, 35.

78 ibid., 37.

79 P. Tikhomirov, "Matematicheskiĭ proekt reformy sotsiologii na nachalakh filosofskogo idealizma," *Bogoslovskiĭ vestnik* 1: 2 (1903), 341.

80 Some vague conservative connotations could have been imparted on mathematics at the close of the nineteenth century, before the Moscow "school" had fully evolved its ideology. Due to the fact that the system of classical education was based on classical languages and mathematics, and that this system was introduced in Russia at a late date (1871) and widely regarded as reactionary, mathematics might have acquired something of a reputation of a "conservative" discipline. It is in this function that mathematics appears in V.A. Gringmut's defense of the classical system of education *Nash klassitsizm* (Moskva, 1890). As well as classical languages (for many years Gringmut taught ancient Greek), though to a lesser degree, mathematics is presented here as a discipline firmly belonging to the educational system inculcating religious faith, the loyalty to the Sovereign, and the national spirit (27–29, 33).

81 See also Chapter 1, Section "Bugaev's library."

82 For the contemporary interest towards discrete functions, see S.S. Demidov, "N.V. Bugaev i vozniknovenie moskovskoĭ shkoly teorii funktsiĭ deĭstvitel'nogo peremennogo," *IMI* 29 (1985), 120.

83 Leo Tolstoy, *War and Peace*, trans. Louise and Aylmer Maude, Vol. III (London: Humphrey Milford, Oxford University Press, 1932), 3–4.

84 An interesting contemporary parallel to Tolstoĭ's reasoning on continuity and history is to be found in the preface of Jacob Burckhardt's *Die Kultur der Renaissance in Italien* (1860). Burckhardt was also concerned with the contrast between the "continuity" of history and the "discreteness" of its representation by historians: "Es ist die wesentlichste Schwierigkeit der Kulturgeschichte, daß sie ein grosses, geistiges Kontinuum in einzelne scheinbar oft willkürliche Kategorien zerlegen muß, um es nur irgendwie zur Darstellung zu bringen" (Berlin: Verlag von Th. Knaur Nachf., 1928, 1–2). From this follows that Burckhardt regarded this problem as the one inherent in a historical narrative to be partly overcome by the proper selection of "discrete" elements. Whereas for the writer the dilemma of continuity/discreteness in history offered an incentive for envisaging a fantastical project of historical research, the historian confined himself to a brief formulation of this dilemma. Interestingly, though, the function of this formulation was in part similar, that is, rhetorical: it was not only a statement of a certain real difficulty, but also an excuse of an essential imperfection of historical narrative.

DOI: 10.1057/9781137338280

85 William McGowen Priestley refers to Tolstoĭ in his history of calculus (after evoking names of its inventors Leibniz and Newton; Leibniz's law of continuity is discussed in one of the preceding chapters): *Calculus: A Liberal Art*, 2nd edn (New-York: Springer-Verlag, 1998), 306, 195–198.

86 The full title of Urusov's work runs as follows: *Obzor kampaniĭ 1812–1813 godov. Voenno-matematicheskie zadachi. O zheleznykh dorogakh* [*A Survey of the Campaigns of 1812–1813. The Military-Mathematical Problems. On the Railways*] (Moskva, 1868). On the influence of Urusov on Tolstoĭ, see B.M. Èĭkhenbaum, *Lev Tolstoĭ: issledovaniia. Stat'i* (Sankt-Peterburg: Fakul'tet filologii i iskusstv SPBGU, 2009), 528–548; on the affinities between Urusov's book and the quoted fragment of *War and Peace*, see Èĭkhenbaum, *Lev Tolstoĭ*, 538–539. ÍA. S. Lur'e's refutation of the relevance of Urusov for the understanding of *War and Peace* (*Posle L'va Tolstogo: Istoricheskie vozzreniia Tolstogo i problemy XX veka* (Sankt-Peterburg: Dmitriĭ Bulanin, 1993), 25) was not convincing.

87 Nekrasov, "Moskovskaia filosofsko-matematicheskaia shkola," 6.

88 Èĭkhenbaum, *Lev Tolstoĭ*, 530, 548.

89 ibid., 532.

90 B.M. Èĭkhenbaum cites the Prince's letter mentioning "intrigues" of the Society against him and his decision not to attend its sittings (*Lev Tolstoĭ*, 539).

91 On his sympathies towards Slavophilism, see Èĭkhenbaum, *Lev Tolstoĭ*, 530–533, 541, 543–544. Let us note, however, that it is hardly possible to imagine Urusov in the company of such representatives of the "school" as Nekrasov or Baron Taube. His ideas alienated Urusov from the Moscow Mathematical Society of earlier times; his character would have been incompatible with the spirit of the "school." Whereas Nekrasov's or Taube's odd use of mathematics was partly related to their sycophancy, Urusov's bizarre notions rather remind one of his remarkable physical courage (Èĭkhenbaum, *Lev Tolstoĭ*, 529–530), suggesting intellectual boldness, even if the latter was pushed to the extreme and turned into eccentricity.

92 ORK i R NB MGU, f. 41, op.1, ed. khr. 87, L. 82 ob. – 83. When exactly these lectures were delivered is not clear.

93 Bugaev regarded Lavrov's study among the "remarkable" works of the year 1859 (ORK i R NB MGU, f. 41, op. 1, ed. khr. 256, L. 2).

94 "'There is no arbitrariness in nature' ['Proizvola net v prirode']. As soon as this thought had grown into a conviction, all the sorcerers, shamans, magicians were rendered impossible; all the gods of the antiquity became nothing <…>. All this scaffolding, with which man had been erecting the temple of the understanding of nature, could be destructed at once, when the cupola of the immutable law emerged before the eyes of humankind <…>" (P. Lavrov, "Mekhanicheskaia teoriia mira," *OZ* 123 (1859), 457).

DOI: 10.1057/9781137338280

95   Mikh. Ferd. Taube, *Sovremennyĭ spiritizm i mistitsizm* (Petrograd, 1909),
       40–41. In particular, Taube argued against the deterministic nature of
       the conceptions of spiritists and insisted on the Creator "being free to
       work miracles" (*Sovremennyĭ spiritizm i mistitsizm*, 24). He identified
       the "*discreteness* and freedom of the creation [*preryvnost'* tvoreniia i
       ego svobodu]" with the "miracle in nature [chudo v prirode]" (Taube,
       Sovremennyĭ spiritizm i mistitsizm, 41; italics are by Taube), and presented
       arithmology as, among other things, aimed at the mathematical explication
       of theological notions (Taube, *Sovremennyĭ spiritizm i mistitsizm*, 218). Cf.
       Nekrasov, "Moskovskaia filosofsko-matematicheskaia shkola," 70; V.A.
       Kozhevnikov, *Sovremennoe nauchnoe neverie. Ego rost, vliianie i peremena
       otnoshenii k nemu* (Sergiev Posad, 1912), 122, 129.

96   Lev Tolstoĭ, *Polnoe sobranie sochineniĭ*, T. 49 (Moskva: GIKhL, 1952), 94; cf.
       Koliagin and Savvina, *Matematiki-pedagogi Rossii*, 58.

97   The letter was addressed to V.G. Chertkov (Lev Tolstoĭ, *Polnoe sobranie
       sochineniĭ*, T. 85 (Moskva: GIKhL, 1935), 60–61).

98   The president of the Moscow Psychological Society, L.M. Lopatin, in his
       speech commemorating Bugaev, mentioned the latter's hope that the
       development of arithmology and neighboring mathematical disciplines
       may bring about "the transformation of the very notion of natural laws"
       (*VFiP* 70: 5 (1903), 900).

99   Seneta, "Statistical Regularity and Free Will," 325–329. For the critique of
       this work by L. von Bortkiewicz, see P.A. Nekrasov, *Theory of Probability*,
       55–64.

100   P.A. Nekrasov, "Filosofiia i logika nauki o massovykh proiavleniiakh
       chelovecheskoi deiatel'nosti (Peresmotr osnovanii sotsial'noi fiziki
       Quetelet)," *MS* 23: 3 (1902) 554–556.

101   ibid., 555 (italics are by Nekrasov).

102   Leo Tolstoy, *War and Peace*, Vol. II (London: Humphrey Milford, Oxford
       University Press, 1931), 258.

103   Within the framework of ultra conservative ideology, this does not seem
       surprising. One may refer to the appeal made by V.A. Gringmut at the first
       All-Russian Congress of the right-wing organizations (February 1906): "Let
       us set free the Double-Headed Eagle, let us spread its wings, let us give it
       back the freedom of its stately flight [vozvratim emu svobodu velichavogo
       poleta]! From now on we shall tirelessly proclaim our demands: 'Freedom
       to the Tsar, Freedom to the Tsar, Absolute freedom to Him! [Svobodu
       Tsariu, Svobodu Tsariu, Neogranichennuiu Emu svobodu!]'" (*Vladimir
       Andreevich Gringmut. Ocherk ego zhizni i deiatel'nosti* (Moskva, 1913), 91).

104   Nekrasov, "Moskovskaia filosofsko-matematicheskaia shkola," 14.

105   ibid., 153 (italics mine).

DOI: 10.1057/9781137338280

106 See R.Sh. Ganelin, *Rossiĭskoe samoderzhavie v 1905 g. Reformy i revoliutsiia* (Sankt-Peterburg: Nauka, S.-Peterburgskoe otdelenie, 1991), 5–6.

107 N.V. Bugaev, *O svobode voli* (Moskva, 1889), 19. For the context of this work, see Shaposhnikov, "Filosofskie vzgliady N.V. Bugaeva," 67.

108 Nekrasov, "Moskovskaia filosofsko-matematicheskaia shkola," 157 (italics are by Nekrasov).

109 ibid., 13.

110 ibid., 154.

111 On the other hand, states "formally governed by the rule of law" are said to substitute "political wisdom" of autocratic monarchy with "irrational legal idols [irratsional'nymi iuridicheskimi kumirami]" (Nekrasov, "Moskovskaia filosofsko-matematicheskaia shkola," 117). The metaphysical framework of such states is the "analytical worldview" with its stress on the "fatal" immutability of laws encouraging the worship of "social idols," such as egalitarianism (see, for example, Nekrasov, "Moskovskaia filosofsko-matematicheskaia shkola," 135).

DOI: 10.1057/9781137338280

3

P. A. Nekrasov: *Theory of Probability* (1912)

Abstract: *Right-wing propaganda responding to the political events of 1911 provided important context to both Belyï's* Petersburg, *and Nekrasov's* Theory of Probability *(1912). In this work Nekrasov presented theory of probability as a powerful economical and, above all, ideological tool capable of defending Russia against the Jews.*

Svetlikova, Ilona. *The Moscow Pythagoreans: Mathematics, Mysticism, and Anti-Semitism in Russian Symbolism.* New York: Palgrave Macmillan, 2013. DOI: 10.1057/9781137338280.

DOI: 10.1057/9781137338280

Introduction

1

In 1905, shortly after the *Moscow Philosophic-Mathematical School* came out, Nekrasov was appointed to the Ministry of Public Education and had to leave Moscow for St. Petersburg. He therefore quit both as warden of Moscow School District and as president of the Moscow Mathematical Society. His writings became increasingly impenetrable, a couple of titles suffice to convey their style: *Gosudarstvo i Akademiîa. Sintez (slozhenie) avtoritetnykh suzhdeniĭ dobrosovestnogo men'shenstva s mneniîami moral'nykh sil doblestnogo bol'shinstva* ([*The State and the Academy. The Synthesis (Addition) of the Authoritative Judgment of the Good-Faith Minority with the Opinions of the Moral Powers* [sic!] *of the Valorous Majority*], 1905); *Vera. Znanie. Opyt. Osnovnoĭ metod obshchestvennykh i estestvennykh nauk (Gnoseologicheskiĭ i nomograficheskiĭ ocherk)* ([*Faith. Knowledge. Experience. The Main Method of the Social and Natural Sciences (An Epistemological and Nomographical Essay)*], 1912).

Historians of mathematics assert that Nekrasov's works written after the beginning of the twentieth century, for all their muddling philosophy, contain much value in them,[1] an example of how profound thoughts can neighbor the darkest follies. Here I shall concentrate on follies, as expressed in a particularly striking form in Nekrasov's *Theory of Probability* (1912).[2] Although it is known that this work concerns ideology, the latter has never been the subject of historical enquiry. It is, therefore, necessary to emphasize that Nekrasov's book must be considered not only within the context of the history of mathematics. It is also, or rather primarily a document of the history of ideology.

2

Theory of Probability has a lengthy preface as obscure as Nekrasov's other works of his later period. It is impossible to give a coherent account of what he writes, but one can give some idea of how his view of the role of mathematics evolved.

Just as in his earlier works, Nekrasov does not speak of universal mathematics, but he undoubtedly has in mind an idea of mathematics as both the principle of the universe (dutifully evoking "Pythagoras" on the world being governed by the number[3]), and the fundamental principle of all human knowledge. This especially concerns probability theory

DOI: 10.1057/9781137338280

which (we are told) "became the basic exact science, hooking onto <...> not only all the fields of the rational knowledge, but of the irrational knowledge, that is of doubt, divination and certainty [stala tochnoîu osnovnoîu naukoîu, zatseplîaîushcheîu <...> vse voobshche oblasti ne tol'ko ratsional'nogo znaniîa, no i somneniîa, gadaniîa i uverennosti ili irratsional'nogo znaniîa]."[4]

This is an elaboration of Laplace's assertion that the "whole system of human knowledge" is related [se rattache] to theory of probability.[5] Passed through Nekrasov's imagination, this thesis becomes associated with the so-called whole knowledge [tsel'noe znanie] of the Slavophils.[6]

The "whole knowledge," intended as a counterbalance to "narrow" and too "rational" Western thought, aspired to embrace not only science but also religious and life experience. In the above-cited passage Nekrasov defines theory of probability as the fundamental discipline, comprising not only rational but also irrational knowledge. Combined with his views on the theological relevance of theory of probability, this gave sufficient reason to integrate the Slavophils into Nekrasov's mathematical mindset, thus yoking together Laplace and Khomîakov.

Another significant feature of the preface is its discussion of the special "probabilistic perception." This is also an echo of Laplace's *Introduction* to *Théorie des probabilités* in which theory of probability is said to be the mathematical expression of a certain "instinct" of "fair minds."[7] Striving to make the esteemed psychologism part of his universal mathematical doctrine, Nekrasov transforms this "instinct" into a new kind of perception placed at the very center of the psyche, and responsible for the most elevated operations of the soul.[8]

Since it is the pure and the immaterial that are associated with the "probabilistic perception," the latter is identified with the Christian "temple of consciousness." Moreover, the "probabilistic perception" is claimed to be the soul itself. It is said to correspond to the "real," which designates the soul in Herbart's psychology.[9] Whether one has or has not this kind of perception distinguishes the living soul from that "buried alive."[10]

All these deliberations, making theory of probability conformable to the Russian philosophy represented by Slavophils, and to Russian Orthodoxy, were, among other things, aimed at demonstrating that theory of probability could become a school discipline of major ideological significance.

From the first page we learn that this edition of *Theory of Probability* was financed by the Ministry of Public Education. Nekrasov expresses

DOI: 10.1057/9781137338280

his deepest gratitude to the Minister of Public Education L. A. Kasso whose name is printed in capitals and invites attention: his educational politics had led to numerous resignations of university professors just the previous year.[11] The way Nekrasov introduces his gratitude is also meaningful:

> The methods of ENLIGHTENMENT [printed in the same way as the name of Kasso below] since olden times have included as one of their most important educational and critical implements the mathematical theory of probability in connection with statistics [PROSVETITEL'NYE metody izdavna schitaĭut odnim iz svoikh vazhneĭshikh obrazovatel'nykh i kriticheskikh orudiĭ teoriĭu veroĭatnosteĭ, v svĭazi so statistikoĭ].

The literal translation of the Russian "Ministry of Public Education" is "Ministry of Public Enlightenment." Nekrasov is trying to represent his work as an effective tool of the educational policy of the present ministry. His bizarre notions were actually meant to be of immediate relevance to the contemporary context. That this was not just idle scheming detached from practice is evident from the very existence of the book—supported by an allowance from the ministry.

3

As a functionary, Nekrasov was deeply involved in the educational reforms launched at the end of the nineteenth century. They were orientated towards replacing the domination of classical education by that of a more practical Realschule type and were therefore favorable to exact sciences. An attempt to make theory of probability a school discipline was partly justified by these circumstances.[12] Moreover, Nekrasov based such projects on very good authority. His immediate source of inspiration were the final words of Laplace's *Introduction* to *Théorie des probabilités*. After having pointed out various advantages arising from knowledge of probability theory, Laplace observed that there was no science more worthy of introduction into the system of public education.[13]

Laplace had nothing to do with Nekrasov's ideological concerns, but the same context of educational reforms that made relevant Laplace's words about the educational value of theory of probability, provides an additional explanation of Nekrasov's involvement with ideology. One of the frequent accusations against classical education was that it nurtured politically indifferent or even revolutionary-minded individuals. To prove the usefulness of a particular school subject it was appropriate

DOI: 10.1057/9781137338280

to stress its ideological relevance.[14] Although Nekrasov seems to have genuinely believed that mathematics could strengthen the right political views, he may have been less obsessive in expressing this belief if the historical situation had been different. The idea of the theory of probability as a school discipline intended to foster Russian Orthodoxy, patriotism and loyalty may thus have appeared less fantastic in the context of the debates around educational reform and, in particular, the ideological aspects of education.[15]

"Dead soul of the world"

1

One of the most striking features of the preface to *Theory of Probability* is the polemics with the then popular novelist Dmitriĭ Merezhkovskiĭ (1866–1941) and the latter's refutation of the principle of the Christian state. The following passage may serve as a good introduction to the theme that dominated Nekrasov's thought at that time:

> The catechism of the mystical word, beginning with the religious-political dogma of the Adventists-Decembrists (see: Member of the Academy D. Merezhkovskiĭ's "Not peace, but sword" <...>), is trying to place in the minds of the masses the notion conceived by the dead soul of the world (the historical noumenal Beast); it [the catechism] gives the name of the Beast to the Christian State, and not to the universal magic dictatorship of the hydra, of the world octopus, enslaving peoples on economical ground. <...> This strange suicidal dogma is purely mystical, Girondist-anarchistic; promising much, it is paving <...> the path to dictatorship and autocracy for the protégés of the dead soul of the world; it is preparing the ground for the power [of the dead soul of the world], and not for the power of the true sovereignty of the crown and the people; by its false hosanna and curse it is preparing in our rear the defeat of our fatherland; it is preparing for us the fate of India, Persia, China, Turkey, where it is in fact the heads of the slippery hydra that are in command.[16]

A little further on, Nekrasov writes of the "magical power of the octopus, of the hydra symbolically represented on the State Emblem at the feet of St. George the Victorious,"[17] the image to be reiterated at the end of *Theory of Probability*, where, after having addressed the subject of national economic growth, the "world Octopus," the "symbolic hydra" "prostrated at the feet of St. George the Victorious"[18] are

DOI: 10.1057/9781137338280

evoked as the promise of a final victory. These words are of crucial importance for the whole of Nekrasov's message, to be clarified in the following pages. By way of anticipation it may, however, be noted that the "symbolic hydra" alluded not so much to the Russian State Emblem, as to the interpretation of the latter by the Black Hundred, according to which St. George slaying the Dragon represented Russia slaying the Jews.

2

As has been mentioned, Nekrasov's teacher Bugaev was extremely anti-Semitic. However, there seems to be no trace of it in his published works. This probably means that there was no important link between his views on mathematics and his anti-Semitism. On the other hand, the archive documents show that the latter was connected with his political conservatism. The usual combination of monarchism and anti-Semitism was shared by both Bugaev, who, as we have seen, marked the Jews in his list of the revolutionary students, and Nekrasov, for whom this list was drawn up. What Nekrasov did was in a way quite logical: he made an attempt to develop Bugaev's ideas of the universality of mathematics by making political ideology an integral and defining part of his mathematical philosophy, with anti-Semitism becoming one of the dominant features of the resultant whole.

In Nekrasov's earlier works there are allusions to the Jews as the source of social disorder. The most remarkable one occurs in the *Moscow Philosophic-Mathematical School* in which there is a passage concerning *"the state within the state*, uniting negative forces for purposes inimical to the state." The state is advised to defend itself against this "treacherous" and "perfidious" adversary, who is called the "political anti-logos," "as if against an enemy who is outlawed, by special ways, suspending the application of the usual principles of law."[19]

By placing this obscure pronouncement in its proper historical setting we detect in *Matematicheskiĭ Sbornik* a justification for pogroms, echoing the rightist propaganda of the previous year, roused by the Kishinev pogrom and the publication of the *Protocols of the Elders of Zion* in *Znamia*.

Yet, the place that the Jews occupy in *Theory of Probability* is different. They are identified with no less than the "dead soul of the world" which must refer to Satan's cunning as the opposite of Sophia, the world soul, God's wisdom and the principle of life.[20]

DOI: 10.1057/9781137338280

Theory of probability, helping to develop the "probabilistic perception" that distinguishes living souls from those "buried alive,"[21] was intended as a Christian weapon against the power of the "dead soul of the world." It is not clear whether Nekrasov thought that the Jews were incapable of learning theory of probability, since they were not Christian and there was no "temple of consciousness" in their souls, or whether he had some racial considerations upon the matter.[22] His mediations were extremely incoherent; there is no doubt, however, that *Theory of Probability* must be regarded within the framework of the contemporary ideology of ultra-Right.

*   *   *

Let us now turn to the political context of Nekrasov's work, and at the same time to the senator of Belyĭ's novel, because it is out of this same historical context that some of the most significant features of *Petersburg* emerged. Belyĭ started to work on the novel in the autumn of 1911. *Theory of Probability* by his father's disciple came out the following year. Both the novel and the mathematical book responded to the same political atmosphere.

1911 in the history of the extreme Right

The year 1911 in Russia may seem a comparatively uneventful one, but only because the history of this period has mainly been written from points of view different to those of the political parties feverishly active in 1911. For the extreme Right, however, 1911 was a crucial year, and it is no coincidence that a recently published collection of documents for the history of the Russian Right contains a large number dating from 1911.[23]

The year 1911 saw a new edition of the *Protocols of the Elders of Zion* published,[24] the anniversary of the assassination of Alexander II—which contributed to the writing of the *Protocols*[25]—fell on March 1, while February 19 had marked the 50th anniversary of Alexander II's abolition of serfdom. The logical scheme employed in the *Protocols* was pervasive in the speeches of the ultra-conservatives delivered on March 1, 1911: the Tsar Liberator was killed by the Jews who seek to enslave the Russians, and the only salvation may come from a strong monarchy.[26]

Further developments in 1911 appeared to substantiate the *Protocols* in the minds of ultra-conservatives.

DOI: 10.1057/9781137338280

In March 1911, a 12-year-old boy, Andreĭ Îushchinskiĭ, was murdered in Kiev. Rumors of ritual murders spread, "and there emerged the question of the Jews consuming Christian blood." This cold and seemingly objective remark is found in the introduction to a masterpiece by Belyĭ's friend Alexander Blok, *Vozmezdie* [*The Retribution*],[27] a conspicuous part of which introduction related to the year 1911.[28] Blok was as anti-Semitic as Belyĭ and was bound to be stirred by the events of that year. This particular one led up to the Beĭlis affair.

Finally, in September 1911, Prime Minister Petr Stolypin was assassinated by Dmitriĭ Bogrov, a Jew who worked for the Okhrana. This produced a very strong impression on the extreme Right:

> <...> the Police Department is full of Jews, and serves, it would seem, the Jews much more than the Russians.

> The recent assassination of Stolypin, arranged by this Department, shows its power. Poor Stolypin had probably participated in the preparation, without even knowing who the victim would be.[29]

This gloomy picture that we find in a letter of the famous Slavist Alekseĭ Ivanovich Sobolevskiĭ (1856–1929) haunted not only his imagination. The Prime Minister's assassination was interpreted as proof of the country being ruled by the Police Department,[30] and the latter, as one could infer from the cases of Bogrov and Azef, was believed controlled by the Jews. The whole situation was repeatedly compared with the Tatar-Mongol yoke, with the only difference being that the Jewish yoke was harsher than that of the Golden Horde.[31]

The future was dark. The overwhelming impression among the ultra Right was that the revolution was at hand or even already happening:

> It is clear that the incident of Stolypin [...] is not a separate episode, but a proclamation of an entire program for the renewal of terror. And this attempt [on Stolypin's life] was only a formal declaration of war, just as was the shooting with live ammunition at the ice hole in front of the Imperial Palace during water consecration on Epiphany Day in 1905 [I èto pokushenie est' tol'ko formal'noe ob"îavlenie voĭny, kak v 905 r. bylo strelîanie kartech'îu po Iordanu Tsarskomu].[32]

The worst misgivings expressed by the *Protocols* seemed to be in fulfillment. It ought to be added that although the anxiety of many conservatives was genuine, the strong emphasis on the pending danger was of particular importance before the election to the Duma to be held the following year.

DOI: 10.1057/9781137338280

The fact that some literary masterpieces created at that time bear distinct traces of the hysterical atmosphere aroused by rightist propaganda makes us look at the latter with close attention. I shall give a characteristic example of this propaganda, which was forgotten afterwards and its literary echoes (including those in Belyï) either misinterpreted, or missed entirely.

Saint-Pierre

L'épouvantable et irréparable malheur qui a subitement anéanti, le 8 mai 1902, Saint-Pierre de la Martinique et dévasté l'une de nos plus riantes colonies, ne nous frappe pas seulement par les deuils et les ruines qui ont été la conséquence immédiate du formidable cataclysme, mais encore par le problème inquiétant qui nous est posé, de connaître les causes réelles de ces éruptions inattendues et de savoir si notre planète est vraiment arrivée à la phase de stabilité sur laquelle nous avons l'habitude d'endormir nos espérances. <...> le 8 mai, jour de l'Ascension, après d'assez graves symptômes prémonitoires, qui n'avaient alarmaient qu'une partie de la population et qu'un grand nombre d'habitants croyaient même à peu près epuisés, une effroyable trombe de gaz enflamé et de produits volcaniques fut lancé du volcan sur la ville, le port et la rade. <...> Toutes la ville s'incendia et s'écroula, les ruines ensevelissant tout. Il semble que de 8 heures du matin jusqu'au soir, Saint-Pierre n'ait été qu'un giganteque brasier. Une immense nappe de feu planait sur toute la région, interdisant toute fuite, brûlant et asphyxiant tous les êtres vivants. Le nombre des victimes dépasse trente mille.

Quelles sont les forces qui ont été en œuvre dans cette révolution du sol?[33]

Camille Flammarion wrote this in the immediate aftermath of the tragedy. The catastrophe which had befallen Martinique in 1902 was apparently of no relevance to what was going on in Russia ten years later. Yet, its place in the political discourse of that time was not negligible. For all the distance separating St. Pierre and St. Petersburg, the similarity of their names was enough to suggest that the destruction of the former had presaged the fate of the latter, and thereby the fate of the Russian Empire. The question about the forces that had provoked the "revolution of the earth" in Martinique was of profound concern for some in Russia in 1911, when the expectation of approaching revolution was abroad. To understand the full significance of the comparison between St. Petersburg, the capital of a state undergoing liberal changes, and Saint-Pierre, the main

DOI: 10.1057/9781137338280

city of a distant French colony, one should bear in mind the reputation of Saint-Pierre. This thriving commercial and industrial center, "amie des lettres, des sciences, des beaux-arts, du progrès"[34] was held to be the Paris of the Antilles, faithful to the republican ideals of its mother-country. All these circumstances, combined with the fact that the tragedy took place on Ascension Day, were bound to attract the attention of those ultra conservatives who exploited popular superstitions (not infrequently sharing them at the same time).

A leaflet issued by a Yaroslavl branch of the Union of the Russians in the beginning of the 1911 may serve to illustrate this. The leaflet is addressed to "brethren of the Union [brat'ia-soîuzniki]" and predicts the end of the world, of which numerous portents are cited. The mention of the mosque being constructed in St. Petersburg ("the city of Saint Peter [grad svîatogo Petra]") suggests the memory of its namesake's fate:

> The Lord has already poured out His righteous ire on the remote city of Saint Peter in the island of Martinique. It had sheltered the Mason temple of Satan [masonskoe kapishche satany], and God exterminated it with fire. There is information that in the Russian city of Saint Peter the cult of Satan has already been established, and that the diabolical "black masses" are performed there; it is terrible to think what is awaiting our unfortunate capital...[35]

Here one must make a pause to remark that the imagination of "brethren of the Union" did not have to work this up entirely on its own. The reader of contemporary publications on the volcanic eruption in Martinique would have been familiar with rumors about the black mass that had provoked the catastrophe.[36] It should also be added that Saint-Pierre was no exception among other flourishing cities in having an active Masonry.[37] With the kind of imagination at work behind the quoted passage, the similarity of names would have been sufficient reason to turn the fate of Saint-Pierre into a moral lesson for Russian readers. As it were, the force of the argument did not rest merely on philological deductions, but drew upon facts and rumors circulating after the tragedy.

Another dreadful natural disaster, the earthquake in Messina in 1909, moved the pious publisher of the *Protocols* to express similar thoughts and fears. Evoking frightful deeds of Masons and Satanists who had provoked God's wrath in Italy, Nilus inquired whether "our once saintly Russia had not taken the same path as this country of hot sun, ripening oranges and lemons [Ne po tomu zhe li puti, chto i èta strana goriachego solntsa, zreîushchikh apel'sinov i limonov, poshla nasha kogda-to Svîataîa Rus'?]"[38] This is a fine specimen of the familiar combination of absurdity

DOI: 10.1057/9781137338280

and staleness of the idiom, characteristic of the sources which nourished Nekrasov's thoughts and style. Saint-Pierre as a gloomy foreboding of future disasters had its place in Nilus' meditations over the future of "once saintly Russia,"[39] all the more so that, like some Bugaev's disciples, Nilus attached enormous importance to the similarity of words. Thus the same book contained an entire diatribe against political and religious movements bearing the "pan-" prefix, which according to the author was an allusion to the pagan god whom Christ's enemies want to revive.[40]

The last section of Nilus' *Bliz grîadushchiĭ Antikhrist i Tsarstvo Diavola na zemle* (1911), which contained the new edition of the *Protocols*, alluded to Martinique and Saint-Pierre, "the capital of Freemasonry" in a discussion of the end of the world.[41]

This line of right-wing propaganda culminated the same year in the appearance of a novel, Elizaveta Shabel'skaîa's *Satanisty XX veka* [*The Satanists of the Twentieth Century*], a sort of literary summary of rightist clichés. Its main characters are Jews and Masons busily engaged in various criminal activities, in particular those of promoting liberal reforms. The final part of the novel unfolds in Saint-Pierre, where the secret Masonic government of the world holds its assembly just before the catastrophe of 1902. The reader is left in no doubt as to the causes of the latter.[42]

1911 in *Petersburg*

The historical context, briefly outlined above, is of defining importance for *Petersburg*.[43]

Belyĭ started writing his novel shortly after Stolypin was killed. He was writing about the revolution of 1905, which the Right believed to have recommenced. A Jew resembling a Mongol is the main driving force of the plot to assassinate Senator Ableukhov. As a pretend revolutionary, the Jew is manipulating an authentic terrorist; as a provocateur, attached to the Okhrana, he is, as we may infer, also manipulating the latter. All these motives were suggested by the right-wing interpretation of the events of the autumn of 1911.

There is also a curious echo of the mood dominating the extreme Right in the beginning of the same year. The demonic Jew of Belyĭ's novel, who does not take money from Okhrana for his service and whose ultimate aims are obscure, bears the name of Lipenskiĭ. It is at the *Lipetskiĭ S'ezd Narodnoĭ Voli* (Congress of People's Will in Lipetsk) that Alexander II was

DOI: 10.1057/9781137338280

sentenced to death. Earlier in 1911 this fact was recalled in a speech by one of the most notable leaders of the extreme Right, Markov II. He described the Lipetsk Congress as a sort of Walpurgis Night, where the Jews decided the fates of the states and monarchs. This line of thought is perfectly represented by Lipenskiĭ in Belyĭ's novel. The name of the character suggested that he was not acting alone, but was a missionary of some dark forces once assembled at the fatal Congress of the similar name.[44]

All this makes one pay closer attention to the image of the apocalyptic earthquake that haunts the terrorist Dudkin. Having an icon of Seraphim of Sarov in his room,[45] he believes that the revolution is bound to grow into an apocalyptic catastrophe. In representing the latter his imagination runs along the traditional lines, those of a great earthquake:

> There will be a leap across history. Great shall be the turmoil. The earth shall be cleft. The very mountains shall be thrown down by the cataclysmic earthquake [ot velikogo trusa], and because of that earthquake our native plains will everywhere come forth humped [a rodnye ravniny ot velikogo trusa izoĭdut povsĭudu gorbom]. Nizhny, Vladimir, and Uglich will find themselves on humps.

> As for Petersburg, it will go down [opustitsa].[46]

For somebody aware of the contemporaneous rumors and extremist writings, the disappearance of St. Petersburg, otherwise natural prey for a deluge, in the wake of an earthquake was quite logical. It appears to have rested upon the interpretation of the end of Saint-Pierre. If by no stretch of imagination one could destroy Saint-Petersburg by a volcanic eruption,[47] an earthquake—part of the disaster in Martinique—was an easier instrument for inflicting God's punishment on the Russian capital.[48]

It should also be remembered that despite his reverence for Seraphim of Sarov, Dudkin at one point becomes a Satanist, and remains one against his will. He thus exemplifies the very sins—involvement in revolutionary activity and Satanism—for which Martinique was punished by way of an admonishment to St. Petersburg.

1911 in *Theory of Probability*

1

In some of Nekrasov's most obscure passages one hears the echo of the latest political developments. Thus, evoking "vampires," "larvae longing

DOI: 10.1057/9781137338280

for blood," and "people's pseudo-laywers" as forces undermining the state,[49] Nekrasov was not simply influenced by conventional anti-Semitic rhetoric, but also by the Beĭlis affair.

The "symbolic hydra" under the feet of St. George the Victorious, traditionally associated with clearing the world of evil enchantment, is magical. Its "dictatorship" is exercised through "witches" and "sorcerers."[50] This reflects a common concern of the extreme Right at the spread of Freemasonry, together with various unorthodox trends regarded within this context as secretly inspired by Freemasonry. Most instructive in this respect are works by Nekrasov's friend Baron M. F. Taube, whose views will be studied in the next chapter. He is likely to have been Nekrasov's immediate source on occult matters. In his review of Nekrasov's book *Matematicheskaia statistika, khoziaĭstvennoe pravo i finansovye oboroty* ([*Mathematical Statistics, Economic Right, and Financial Turnovers*], 1910), published in 1911, there is an observation on the close link between modern economics and occultism.[51] Another of Taube's works of the same year provides us with a clue to a better understanding of the "dead soul of the world." Amidst an invective against black magic, he speaks of the "collective sepulchral social soul" (in Russian it is no less absurd: "kollektivnaia zamogil'naia dusha obshchestva") invoked by occultists for their satanic ends.[52]

Seen against this background, the "dead soul of the world" signified not only the anti-Christian principle, personified by the Jews, but also the most dreadful instrument they were supposed to use. In the *Moscow Philosophic-Mathematical School* we detected traces of Nekrasov's familiarity with some sources on psychic phenomena, which constituted a natural sphere of interest for a pupil of Bugaev and a member of the Moscow Psychological Society. To grasp the extent to which the ideology proposed in *Theory of Probability* was really obscurantist, it is important to realize that the author's earlier views had now evolved to embrace belief in black magic.

2

The "symbolic hydra" is regularly quoted in the studies on Nekrasov but no adequate commentaries are offered. One of the most important elements of the ideological context of *Theory of Probability* appears to be entirely forgotten.[53] "Symbolic hydra" is an unmistakable allusion to the *Protocols*, in which "symbolic snake" is the subject of a special discussion: "<...> our goal is now almost a few steps off. There remains a small space to cross,

DOI: 10.1057/9781137338280

and the whole of the path we have trodden is ready to close its cycle of the *Symbolic Snake*, by which we represent our people. When this ring closes, all the European states will be locked in its coils, as in a powerful vice."[54] In the notes accompanying the protocols, this statement is furnished with further details: "The sages decided by peaceful means to conquer the world for Zion with the slyness of the Symbolic Snake, whose head was to consist of the Government of the Israelites initiated in the plans of the sages <…>."[55] Seemingly, "heads of the slippery hydra" derive from this.[56]

The warning quoted against the "fate of India, Persia, China, Turkey" ruled by the "hydra" partly refers to the stages of the Jewish conquest of the world described in the *Protocols*: Constantinople is the last city separating the "Snake" from its aim, Jerusalem, upon reaching of which it will complete its embrace. In the 1911 edition, Nilus exclaims that "Constantinople has already been captured."[57]

On the other hand, the names of "India, Persia, China, Turkey" reflect anxieties aroused by real political events. The National Congress movement in India had been growing in intensity and violence since 1910. Persia had been passing through revolution since 1905. China's revolution of 1911 led up to the establishment of the Republic of China.[58] Nilus' exclamation referred to the Turkish revolution (started in 1906). Within the conceptual framework of the *Protocols*, all this turmoil bore witness to the Jewish conspiracy.

The "symbolic Golden Calf," one of Nekrasov's synonyms of the "dead soul of the world," is another allusion to the same source and its various commentaries:

> The task is clearly expressed: the enslavement of all humankind by the tribe of Judah, along with the smashing of the Cross and the Christian Faith and their replacement by the Reign of the Golden Serpent. The Serpent is to be identified with the tribe of Judas, and with financial capital, and with the deified Golden Calf <…>.[59]

3

A mathematician believing in the authenticity of the *Protocols* must have paid special attention to the protocols 8 and 20 in which the "question of figures [vopros tsifr]," that is, economics, is evoked as decisive. It may be argued that *Theory of Probability* is intended to provide a weapon against the Jews, represented in contemporaneous propaganda as threatening universal economic slavery. These are the historical meanings aggregated in the image of St. George slaying the World Octopus in Nekrasov's book.

DOI: 10.1057/9781137338280

Theory of Probability is one of the most interesting manifestations of the idea of the Jewish threat. Curiously, however, Nekrasov does not mention the word "Jew." Only the "Rothschilds" make a brief appearance in the end.[60] Nevertheless, if there are any doubts as to the actual meaning of Nekrasov's references to the "hydra" and the like, the cited collection of documents on the history of the right-wing parties dispels them.

In this collection we discover the proceedings of the fifth All-Russian Congress of the Russian People [russkikh lîudeĭ] that took place at the same time as the fourth All-Russian Congress of the Union of the Russian Nation [russkogo naroda], in May 1912. Nekrasov participated in the former. The proceedings record his proposition to create the Lomonosov society, one of whose functions would be to protect the Russian language.[61] This strikes one as a remarkable proposition, since few could match Nekrasov in producing unintelligible Russian. On the other hand, this shows that his style of writing had an obscure ideological ground. Apparently, he gave thought to how he wrote.[62]

In the same speech Nekrasov spoke of measures against the Jews who would try to seize power over the newly organized society. This evidence is precious as one of the rare explicit formulations of his views.[63]

4

Nekrasov sent his *Vera, znanie, opyt* (the abridged popular version of *Theory of Probability*) to one of the leaders of the Black Hundred movement Boris Vladimirovich Nikol'skiĭ (1870–1919) recommending it as "criticizing in mathematical language the *Critique* of the pure and practical reason of Kant [kritikuîushchuîu 'Kritiku' chistogo i prakticheskogo razuma Kanta matematicheskim îazykom]."[64]

From the point of view of the history of ideology, Nekrasov's attitude towards Kantianism adopted in *Theory of Probability* deserves special attention. The following argument is particularly interesting:

> The order, the harmony, the triple correspondence, in the cognoscible, the cognizing, and in the word, represent for consciousness the highest value which is opposed to the Chaos and the Night. Where this subjective-objective catholic [in the Russian Orthodox sense] value has not set in [Gde vovse ne otstoîalos' ètoĭ sub"ectivno-ob"ectivnoĭ sobornoĭ tsennosti], it is impossible to speak about clear foresight, spiritual freedom, heroic exploit, equal rights, love and other values [o îasnom predvidenii, o dukhovnoĭ svobode, o podvige, o ravnom prave, o lîubvi i drugikh tsennostîakh].[65]

DOI: 10.1057/9781137338280

Evoking "the Chaos and the Night," Nekrasov, given the epistemological character of the passage, must have had in mind the preface to the first edition of *The Critique of Pure Reason*. The context in which "the Chaos and the Night" were mentioned by Kant was very different—speaking of the current state of mind among philosophers, he characterized the reigning indifference as "die Mutter des Chaos und der Nacht." However, the very expression with its vague archaic associations would be very much to Nekrasov's taste, so that in formulating his views on the epistemological matters, Kantian words were a likely prompt. Being another illustration of how Nekrasov transformed the modern and the contemporary into archaic, the "Chaos" to which Nekrasov alluded and which hinted at Kant and his followers, who were opposed to the conception of the Christian epistemological "harmony," corresponded to the political framework outlined above.[66] It will be remembered that in *Petersburg*, Ableukhov the younger's studies of Kant and neo-Kantians were presented as directly related to his Semitic origins. Two factors were especially relevant in this connection: that many of the important neo-Kantians were Jewish, and that one of them, Hermann Cohen, established a parallel between Kantianism and Iudaism.[67]

5

The intense atmosphere aroused by the propaganda of the extreme Right forms the historical background to the second edition of *Theory of Probability*. The latter must be regarded as an ideological document that reflects the great commotion among the extreme Right following the events of 1911. Not only are some of the most significant features of Nekrasov's book prompted by this ideological climate, but possibly the very fact of its existence is. Although Nekrasov remembered about the centenary of *Théorie analytique des probabilités* by Laplace,[68] the dating of Nekrasov's book cannot be explained with reference to this anniversary, but should be explained by contemporary events. For years, Nekrasov had had the idea of the ideological use of mathematics in general, and probability theory in particular. Now he would have had some reason to believe that, in the face of the imminent peril, his words would be heard. He believed himself in possession of the most effective remedy against the greatest danger: once implanted in peoples' minds, probability theory would render the Jews powerless.

The very attempt to produce a mathematical book embroiled with philosophy and ideology is better understandable against the background

DOI: 10.1057/9781137338280

of the anti-Semitic tradition, which represented the Jews as naturally predisposed to spinning abstract dogmatic webs[69] entrapping Aryans. Thus combining abstract mathematical notions with ideology could serve as a means of protection against Jewish influence.[70]

On the other hand, the same background helps account for the above mentioned juxtaposition of the "probabilistic" worldview and the "whole knowledge" of Slavophils, whose philosophy was supposed to be not an exercise of abstract thought, but a spiritual practice fused with life. It was now a matter of life and death to strengthen this allegedly national Russian outlook by providing it with a mathematical foundation.[71]

6

Theory of Probability reminds one of other instances of similar intent to create a universal science, and in particular those brought back from oblivion by Frances Yates. While reading Nekrasov's celebrations of harmony in its various forms, one might feel oneself in the presence of an academician delivering a speech in a French academy of the sixteenth century. The religious enthusiasm, the drive for an all-embracing knowledge, and the archaic ring so distinct in some of his phrases seem to be promising something quite different from what we have found. "<…> this lady [mathematics] is the prerequisite of the victory over the evil of world; driving her out means driving out this victory [èta dama est' predposylka pobedy nad zlom mira; gnat' ee znachit gnat' ètu pobedu]."[72] Such exalted frame of mind might be congenial to that of the academicians evoked by Yates. But what in the former case seems to be the side effect of great creative energy, which naturally generated hopes never to be fulfilled—those of universal knowledge and social harmony—in the latter case is suggestive of a disease to become particularly rampant later on in the last century. The striving for universality can betray the propensity for forming idées fixes, which was clearly characteristic of Nekrasov. Anti-Semitism was another facet of the same propensity. Universal mathematics of the type elaborated by the "school" was, in a sense (structurally, so to speak), a natural ally of the most pervasive anti-Semitism.

* * *

Nekrasov was not an isolated thinker. I shall consider three of the authors belonging to the circle of the Moscow "school."[73] The first, V. G. Alekseev, is known, but his pedagogical works have attracted little attention. The

DOI: 10.1057/9781137338280

second, Baron M. F. Taube, is never mentioned in connection with the "school,"[74] though, from Nekrasov's point of view, he was one of its most prominent members. And finally, the third, P. A. Florenskiĭ, is widely known, but some of the aspects of his connections with the ideology of the school require elucidation.

Notes

1   Eugene Seneta, "Statistical Regularity and Free Will," *International Statistical Review* 71: 2 (2003), 331. M.V. Chirikov and O.B. Sheynin wrote that certain Nekrasov's thoughts "enable us to consider him as some mathematical Nostradamus" ("Perepiska P.A. Nekrasova i K.A. Andreeva," *IMI* 35 (1994), 128; http://www.sheynin.de/download/2_Russian%20Papers%20History.pdf, 71). Cf. S.S. Petrova, "Iz istorii prepodavaniia matematiki v moskovskom universitete s 60-kh gg. XIX – do nachala XX veka," *IMI. Vtoraia seriia* 46: 11 (2006), 143–144.

2   The title-page of this book tells us that it is the second edition. Indeed, there was Nekrasov's earlier work of the same title dating back to 1896. The "second edition" is more than double its size and differs from the first to the point of being a totally new book. The above mentioned *Vera. Znanie, Opyt* is an abridged version of *Teoriia veroiatnosteĭ* of 1912.

3   P.A. Nekrasov, *Teoriia veroiatnosteĭ* (S.-Peterburg, 1912), XV.

4   ibid, I.

5   See Chapter 2, n. 9.

6   Nekrasov, *Teoriia veroiatnosteĭ*, II.

7   "<…> la théorie des probabilités n' est, au fond, que le bon sens réduit au calcul; elle fait apprécier avec exactitude ce que les esprit justes sentent par une sorte d'instinct, sans qu'ils puissent souvent se rendre compte" (Laplace, *Oeuvres complètes*. Paris: Gauthier-Villars, Imprimeur-libraire, 1886, CLIII); cf. "Otvet P.A. Nekrasova na zamechaniia i vozrazheniia otnositel'no filosofskikh i logicheskikh osnovaniĭ sotsial'noĭ fiziki," *VFiP* 68: 3 (1903)," 597.

8   Nekrasov, *Teoriia veroiatnosteĭ*, VII–VIII.

9   Cf. V.G. Alekseev, *Plody vospitatel'nogo obucheniia v dukhe Komenskogo, Pestalotstsi i Gerbarta* (ÎUr'ev, 1906), 110.

10 Nekrasov, *Teoriia veroiatnosteĭ*, VII.

11 Cf. Seneta, "Statistical Regularity and Free Will," 325; Vygodskiĭ, "Matematika i ee deiateli v Moskovskom universitete vo vtoroĭ polovine XIX v." *IMI* 1 (1948), 177.

12 See P.A. Nekrasov, ed., *Teoriia veroiatnosteĭ i matematika v sredneĭ shkole* (Petrograd, 1915). In 1909–1910 Nekrasov had already come up with the idea of introducing theory of probability (along with statistics, accounting

DOI: 10.1057/9781137338280

and geography) into the curriculum of the so-called gymnasia with new languages (GARF, f. 1838, op. 1, ed. khr. 3090, L. 14 ob). On Nekrasov's attempts to introduce probability theory into the school curriculum, see, in particular, Oscar B. Sheynin, "Nekrasov's Work on Probability: The Background." *Archive for History of Exact Sciences* 57 (2003), 343–345; A.P. Îushkevich, *Istoriîa matematiki v Rossii do 1917 goda* (Moskva: Nauka, 1968), 311; Vygodskiĭ, "Matematika i ee deîateli," 178. Useful material concerning the context of educational reforms of the beginning of the twentieth century is to be found in V.M. Busev, "Shkol'naîa matematika v systeme obshchego obrazovaniîa 1918–1931 gg.," *IMI. Vtoraîa seriîa* 47: 12 (2007), 68–75.

13   Laplace, *Oeuvres complètes*, CLIII.

14   It was mostly the humanities which were thus discussed, but within Nekrasov's universalistic outlook they were related to mathematics and vice versa. He wrote of mathematics as a "humanistic discipline [gumanitarnaîa nauka]" (*Teoriîa veroîatnosteĭ*, IX), and was very keen on emphasizing its role in fostering "humanism," the whole argument being immediately related to the ongoing intense critique of the classical education (that is the tradition of Humanism).

15   In view of Nekrasov's close reading of Plato, he may have associated his educational ideas with the book VII of the *Republic*. It will be remembered that in Plato mathematics is indispensable not for its practical use, nor merely because it develops abstract thought, but for ideological reasons: it helps to form a frame of mind necessary for the rulers of the ideal state.

16   Nekrasov, *Teoriîa veroîatnosteĭ*, XXVIII. In the cited work Dmitriĭ Merezhkovskiĭ—who, it should be noted, was not elected in the Imperial Academy—called the Russian Empire a "City of the dead" inhabited by Gogol's "dead souls." See D. Merezhkovkiĭ, *Ne mir, no mech. K budushcheĭ kritike khristianstva* (S.-Peterburg: Izd. M.V. Pirozhkova, 1908), 67–69.

17   Nekrasov, *Teoriîa veroîatnosteĭ*, XXXI.

18   ibid., 532.

19   Nekrasov, "Moskovskaîa filosofsko-matematicheskaîa shkola," 136–137. Nekrasov reiterates this thought in his book *Gosudarstvo i Akademiîa* (Moskva, 1905, 34–35), in which "physiological heredity" is claimed to be one of the factors which unite the secret enemies of the state.
Shafarevich's discussion of the subversive activities of the "small people" (see Introduction), bearing evident similarities with this argument of Nekrasov, belongs to the same ideological tradition.

20   At that time the term "world soul" was widespread in Russian mostly owing to Vladimir Solov'ev, one of Nekrasov's influences (see Sheynin, "Nekrasov's Work on Probability," 342; P.A. Nekrasov, *Theory of Probability*, compiled, translated and commented by Oscar Sheynin: http://www.sheynin.de/

DOI: 10.1057/9781137338280

download/5_Nekrasov.pdf, 2). On possible Pythagorean connotations of the "world soul" in this context see Chapter 5, n. 52.

21 See above.

22 Cf. Chapter 5, Section "Apollo."

23 ÎU.I. Kirîanov, ed., *Pravye partii. 1905–1917. Dokumenty i materialy. V 2 tt.*, T. 2, 1911–1917 gg. (Moskva: ROSSPÈN, 1998), 9–88.

24 [Sergeï Nilus], *Bliz grîadushchiï Antikhrist i Tsarstvo Diavola na zemle* ([Sergiev Posad, 1911]), 57–133.

25 Cf. Chapter 1, Section "Anti-Semitism."

26 Kirîanov, *Pravye partii*, T. 2, in particular, 13–15; 19–20.

27 Aleksandr Blok, *Sobranie sochineniï v 8 tt.*, T. 3 (Moskva – Leningrad: GIKhL, 1960), 296. The poet Osip Mandel'shtam was indignant at the phrase and mocked at its objective tone. The notes which he made on the margins of the copy of the *Retribution* which belonged to Èmma Gershtein all started with "and there emerged the question [i voznik vopros]." See Èmma Gerstein, *Vospominaniïa* (S.-Peterburg: INAPRESS, 1998), 17.

28 Blok, *Sobranie sochineniï v 8 tt.*, T. 3, 296–297.

29 [ÎU. I. Kirîanov, ed.], "Perepiska i drugie dokumenty pravykh (1911–1913)," *Voprosy istorii* 10 (1999), 109.

30 Blok, *Sobranie sochineniï v 8 tt.*, T. 3, 297.

31 Kirîanov, ed., *Pravye partii*, T. 2, 206. Cf. Aleksandr Selîaninov, *Taïnaîa sila masonstva* (S.-Peterburg, 1911), 294.

32 This letter by D.A. Khomîakov, an active Rightist and the son of the Slavophil, was written on the September 3, 1911, before the death of Stolypin. See [ÎU. I. Kirîanov, ed.], "Perepiska i drugie dokumenty pravykh (1911 god)," *Voprosy istorii* 11–12 (1998), 127. Khomîakov alludes to the incident which took place on Epiphany day in 1905, when during the spectacular ceremony of water consecration, with the Tsar at the head of the festive procession, one of the cannons which had to salute shot with live ammunition. Though there were no victims (only one policeman was wounded and some windows of the Winter Palace were broken), this incident came to be regarded as a sinister portent of the coming revolution: three days later the so-called Bloody Sunday occurred: a peaceful demonstration of workers was broken up by the army, which triggered the first Russian Revolution.

33 Camille Flammarion, *Les éruptions volcaniques et les tremblements de terre: Krakatoa, la Martinique, Espagne et Italie* (Paris: Ernest Flammarion, Éditeur, 1902), 97–99.

34 Coeur Créole [Ch. L. Lambolez], *1635–1902: Saint-Pierre-Martinique. Annales des Antilles française—journal et album de la Martinique. Naissance, vie et mort de la cité créole. Livre d'or de la charité* (Paris-Nancy: Berger-Levrault & Cⁱᵉ, Éditeurs, 1905), 90.

DOI: 10.1057/9781137338280

35   [ÎU. I. Kirîanov, ed.], "Perepiska i drugie dokumenty pravykh 1911 goda," *Voprosy istorii* 10 (1998), 100.

36   "'Après un banquet sacrilège, le jour du Vendredi-Saint de l'année 1902, des hommes auraient gravi la montagne Pelée et, quand ils eurent procédé à un crucifiement monstrueux, ils auraient jeté dans le cratère de la montagne les débris d'une croix. Le cratère a répondu et ce fut le 8 mai, jour de l'Ascension'" (F. de Croze, *La Martinique. Catastrophe de Saint-Pierre* (Limoges: Marc Barbou, Éditeur, 1903), 136; the author introduces the quotation as a "fact" cited by "newspapers [des feuilles publiques]").

37   On the Masons of Martinique, see Gerry L. Prinsen, ed., *The Story of the Lodge La Parfaite Union, in the Island of Martinique* ([Kila, Mont.]: Kessinger Publishing, [1998]; Reprint of [Helmond]: Latomia, 1993); J.-M. Ragon, *Orthodoxie Maçonnique, suivie de la maçonnerie occulte et de l'initiation hermétique* (Paris: E. Dentu, Libraire-Éditeur, 1853), 151.

38   Sergeï Nilus, *Na beregu Bozh'eï reki. Zapiski pravoslavnogo* (Sergiev Posad, 1916), 27. These notes, according to Nilus himself, formed his diary conducted in 1909.

39   Nilus, *Na beregu Bozh'eï reki*, 10.

40   ibid., 85–86. Nilus' ravings may seem worlds away from the refined mysticism of the younger generation of Bugaev's pupils as depicted in the book by Graham and Kantor. Yet, the somewhat similar attitude towards the power of words, which if taken out of its historical context may suggest a tantalizing poetical frame of mind was not the only superstition shared by Moscow mathematicians with Nilus. In this respect one should mention D.F. Egorov's belief in the famous prophecy of Seraphim of Sarov (which included the frequently quoted words: "Satan was the first revolutionary, through which he fell from heaven"; S.S. Demidov, "Professor Moskovskogo universiteta Dmitriĭ Fedorovich Egorov i imeslavie v Rossii v pervoĭ treti XX stoletiĭa," *IMI. Vtoraîa seriîa* 39: 4 (1999), 146). Egorov showed great courage in defending his religious beliefs. These beliefs, however, for all their possible beneficial impact on mathematical studies, are very likely to have had a deeply obscurantist side to them (the "Prophecy of Seraphim of Sarov recorded by Motovilov" was published by Nilus). What is known about the views of some other Name worshipers, such as Florenskiĭ and A.F. Losev, tends to corroborate this supposition. On Losev, see Michael Hagemeister, "Pavel Florenskij und der Ritualmordvorwurf," in Michael Hagemeister and Torsten Metelka, eds, *Appendix 2. Materialien zu Pavel Florenskij* (Berlin u. Zepernick: Kontext, 2001), 72, n. 38. It is also instructive to read a chapter in Losev's *Esthetics of Renaissance* on Savonarola, wherein the latter is praised as transmitting a message of enlightenment and humanism. See A.F. Losev, *Èstètika vozrozhdeniîa* (Moskva: Mysl', 1978), 567–582.

41   Nilus, *Bliz griadushchiĭ Antikhrist*, 199–201.

DOI: 10.1057/9781137338280

42 E. Shabel'skaîa, *Satanisty XX veka. Roman (Ottiski fel'etonov gaz. "Kolokol" za 1911). Vyp. 1. Ch. I–II* (S.-Peterburg.: Izd. V.M. Skvortsova, 1912). Cf. G.V. Obatnin, "Proteus: eshche raz o satanistakh XX veka," *Russkaîa literatura* 4 (2007), 45.

43 See Mikhail Bezrodnyî, "O 'îudoboîazni' Andreîa Belogo," *Novoe literaturnoe obozrenie* 28 (1997), 100 – 125; Magnus Ljunggren, *Twelve Essays on Andrej Belyj's Peterburg* (Göteborg: Göteborgs Universitet, 2009), 63–71; Ilona Svetlikova, "Kant-semit i Kant-ariets u Belogo," *Novoe literaturnoe obozrenie* 93: 5 (2008), 62–98.

44 Svetlikova, "Kant-semit i Kant-ariets u Belogo", 66–67.

45 Andreî Belyî, *Peterburg* (Sankt-Peterburg: Nauka, 2004), 242.

46 Andrei Bely, *Petersburg* trans. Robert A. Maguire and John E. Malmstad (Bloomington – London: Indiana University Press, 1978), 65.

47 The paintings imitating frescoes of Pompeii in Ableukhovs' apartment (Belyî, *Peterburg*, 87) probably serve as another reminder of the fate of Petersburg, the namesake of which was called "Pompeii of the Antilles."

48 That Belyî must have shared this interpretation of the end of Saint-Pierre is obvious from his later poem *Sovremennikam* (1918), where the downfall of the "sinful Martinique" is evoked. That he viewed the disaster of 1902 as some portent of the future follows from the theses of his lecture, also delivered in 1918, *Svet iz griadushchego* [*The Light from the Future*]. In one of them we read: "The destruction of culture. The future 'Martiniques' [Grîadushchie 'Martiniki']. East or West" (RGB, f. 25, k. 31, ed. khr. 14, L. 1).

49 Nekrasov, *Teoriîa veroîatnosteî*, VI.

50 ibid., VI.

51 M.F. Taube, "Prilozhenie osnovnykh zakonov myshleniîa k voprosam statistiki i finansovogo pravomeriîa," *MT* 5 (1911), 94.

52 M.F. Taube, "Uchenie o pustote, kak osnova budizma," *MT* 8 (1911), 61.

53 There are two particularly interesting examples of such forgetfulness. The following is the quotation from the article on Nekrasov by Sergeî Polovinkin: "The principle <…> of the unification of the state is the Sovereign <…>. Nekrasov represented the inner corrupting forces which oppose this unification as the 'symbolic hydra' on the State Emblem of the Russian Empire, prostrated at the feet of St. George the Victorious. This 'world octopus,' according to his opinion, is the world capital" (S.M. Polovinkin, "Psikho-aritmo-mekhanik (filosofskie cherty portreta P.A. Nekrasova)," *Voprosy istorii estestvoznaniîa i tekhniki* 2 (1994), 111–112). Polovinkin's account of Nekrasov's views is seemingly objective. He does not include any commentaries to the theses he cites, or to the expressions he quotes. As a result we have a sympathetic account of the Black Hundred's position, all the more interesting, that, as mentioned earlier, Polovinkin happens to have also published a book of documents on Nilus.

DOI: 10.1057/9781137338280

Another author on Nekrasov is more explicitly favorable to the Black Hundred: "The reasons for ranging Nekrasov among the 'Black Hundred' [by the Soviet critics of the 'school'] were more than enough. First of all, there is his sincere monarchism; then, naturally, his Russian Orthodoxy. Moreover, through the lines of some of Nekrasov's passages one can even see, if only one wishes [pri zhelanii], an indispensable attribute of the cliché of the 'Black Hundred', that is anti-Semitism" (A.V. Andreev, "Teoreticheskie osnovy doveriia (shtrikhi k portretu P.A. Nekrasova)," *IMI. Vtoraia seriia* 39: 4 (1999), 104). The inference is that anti-Semitism was only superficially (if at all) connected with the Black Hundred, and that Nekrasov, who shared the Black Hundred's allegedly innocuous views, was really not anti-Semitic.

54   Nilus, *Bliz griadushchiĭ Antikhrist*, 66 (italics are by Nilus); cf. *The Protocols of the Meetings of the Learned Elders of Zion*, trans. Victor E. Marsden, http://ddickerson.igc.org/The_Protocols_of_the_Learned_Elders_of_Zion.pdf, 30.

55   Nilus, *Bliz griadushchiĭ Antikhrist*, 134; cf. *The Protocols of the Meetings of the Learned Elders of Zion*, 20.

56   See Section "Dead soul of the world."

57   Nilus, *Bliz griadushchiĭ Antikhrist*, 154.

58   At the beginning of 1911 there was a conflict between Russia and China which revived fears of the "yellow danger" and of the approaching war with China.

59   G. Butmi, *Vragi roda chelovecheskogo*, 4th edn (S.-Peterburg, 1907), 107.

60   Nekrasov, *Teoriia veroiatnosteĭ*, 531.

61   Kirianov, *Pravye partii*, T. 2, 172. 1911 was the anniversary of M.V. Lomonosov. M.F. Taube published a brochure which presented Lomonosov as an inspirer of the Black Hundred movement (Bugaev is mentioned here among the "luminaries" following in the path struck out by Lomonosov). See Mikhail Vashutin [M.F. Taube], *Mikhail Vasil'evich Lomonosov* (Petrograd, 1911), 2.

62   Cf. P.A. Nekrasov, "Logika mudrykh liudeĭ i moral' (Otvet V.A. Gol'tsevu)." *VFiP* 70: 5 (1903), 926–927; id., "Predislovie," *MS* 25: 1 (1904), XIV.

63   Another example is to be found in the English version of Oscar Sheynin's article "Publikatsii A.A. Markova v gazete 'Den'' za 1914–1915 gg." (*IMI* 34 (1993), 194–209): http://www.sheynin.de/download/2_Russian%20Papers%20History.pdf, 63. See also n. 66.

64   GARF, f. 588, op. 2, ed. khr. 102, L. 1. The letter is dated January 28, 1916.

65   Nekrasov, *Teoriia veroiatnosteĭ*, 110.

66   In his second letter to Nikol'skiĭ dated February 16, 1917 Nekrasov bitterly complains of Russian intelligentsia becoming "Jewish [ob"evreivaetsia]" (GARF, f. 588, op. 2, ed. khr. 102, L. 4).

67   Thus, in 1911 the reader of the widely read journal *Russkaia mysl'* was informed that Hermann Cohen was "trying to philosophically interpret and justify the Jewish religion <...> and to show its superiority over

DOI: 10.1057/9781137338280

Christianisty." See S.O. Margolin, "Filosofiia evreĭskoĭ religii," *Russkaia mysl'* 3 (1911), 37 n. 1. It will be recalled that originally Belyĭ was planning to publish *Petersburg* in *Russkaia mysl'* but the latter's editor, P.B. Struve, refused to accept the novel.

68   Nekrasov, *Teoriia veroiatnosteĭ i matematika*, 15.

69   Houston Stewart Chamberlain, *The Foundations of the Nineteenth Century*, trans. John Lees, 2nd edn, Vol. II (London – New York: The Bodley Head; John Lane Company, 1912), 19. For a variation of this theme in *Petersburg*, see Appendix 3.

70   Nekrasov inveighs against the danger of abstract mathematical thought in a later work. See P.A. Nekrasov, *Po povodu stat'i akademika A.A. Markova o proekte prepodavaniia teorii veroiatnosteĭ v sredneĭ shkole* (Petrograd, 1915), 17, 18.

71   Cf. Chapter 4, Section "Baron M.F. Taube." Aryan connotations of the universality of the philosophic-mathematical outlook would be particularly fitting in this context, but they seem to be absent. The probable explanation is that speaking of the Aryans was too closely associated with theosophy which had become increasingly suspicious in the eyes of Russian Orthodoxy (see, for example, Taube, "Uchenie o pustote," 60–61).

72   Nekrasov, *Teoriia veroiatnosteĭ*, XV n.

73   In view of the fact that the "school" was an ideological trend, its membership cannot be established with precision. For a list of works that Florenskiĭ considered as representing the ideas of the "school," see his *Stolp i utverzhdenie istiny* T. 1 (II) (Moskva: Pravda, 1990), 683–684 n. 182.

74   To my knowledge, the only exception is the book by ÎU. M. Koliagin *Matematiki-pedagogi Rossii. Zabytye imena. Kn. 3. Pavel Alekseevich Nekrasov* (Orel: GOU VPO "OGU," OOO "Kartush-PF," 2008), 87, 107, n. XXI.

DOI: 10.1057/9781137338280

4

Some Other Members of the "School"

Abstract: *The works of three other members of the Moscow philosophic-mathematical school—V. G. Alekseev, M. F. Taube, P. A. Florenskiĭ—provide additional insight about the ideas of the "school," and its right-wing political framework.*

Svetlikova, Ilona. *The Moscow Pythagoreans: Mathematics, Mysticism, and Anti-Semitism in Russian Symbolism.* New York: Palgrave Macmillan, 2013.
DOI: 10.1057/9781137338280.

DOI: 10.1057/9781137338280

V. G. Alekseev

Vissarion Grigor'evich Alekseev (1866–1943?[1]) is of interest here because his works show that Nekrasov's ideas about the didactic use of mathematics could be conveyed in a much more reasonable manner and thereby become closer to being put into effect.

On a minor scale Alekseev's career repeated that of Nekrasov. He was a professor of mathematics at Îur'ev (formerly Dorpat) University. He was elected dean of its physical–mathematical faculty, and then its pro rector. In 1909, during a crackdown that followed a period of liberal transformations at the university started in 1905 (numerous clauses against the Jews had been abolished and women permitted as auditors), Alekseev was appointed its rector, probably not without Nekrasov's assistance. The previous rector, who had not been appointed but elected, was removed by the Ministry of Public Education, which launched an investigation against him.[2] Then, in 1914 Alekseev became warden of the Wilno School District, and a year later—a member of the Council of the Ministry of Public Education.[3]

I shall not touch upon Alekseev's philosophical works (influenced by both Bugaev and Nekrasov), but only on the pedagogical ones, in which his ideological involvement is more readily apparent. In particular, his views may be gathered from his two books on education written after 1905: *Plody vospitatel'nogo obucheniia v dukhe Komenskogo, Pestalotstsi i Gerbarta* ([*The Fruits of the Educational Training in the Spirit of Komenský, Pestalozzi and Herbart*], 1906) and *Gerbart, Strumpel i ikh pedagogicheskie sistemy* ([*Herbart, Strümpel and Their Pedagogical Systems*], 1907). Both of them treat mathematics as an ideological discipline to be studied in school for the explicit purpose of appeasing social unrest. Herbart is one of his most frequent references. At the time of the educational reforms in Russia, Herbart's pedagogical works were translated and discussed. There was perhaps some association between Herbartian pedagogy and conservatism, the former being regarded as the basis for harmonious education creating a stable society. We come across the expression of such hopes in an anonymous booklet published immediately after the assassination of Alexander II.[4]

At any rate, it is as part of the conservative education agenda that Herbart appears in Alekseev's works, invoked as an authority on mathematics as a remedy for intellectual ferment: whereas philosophy is a notorious cause of social disturbances, mathematics with its firm and

DOI: 10.1057/9781137338280

clear rules presents an invaluable antidote. Dangers of abstract philosophizing are to be overcome with the help of mathematics.[5] With the memories of the revolution of 1905 still afresh, these ideas were topical: mathematics was presented as a beneficial cure for a troubled society.[6]

In the Russian National Library there is a copy of the first of the cited books (*Plody vospitatel'nogo obucheniia v dukhe Komenskogo, Pestalotstsi i Gerbarta*) with the following inscription: "To Her Imperial Highness the Grand Duchess Militsa Nikolaevna with respect from the author [Ee imperatorskomu Vysochestvu Velikoĭ Knîagine Militse Nikolaevne s pochteniem ot avtora]". The Grand Duchess Militsa Nikolaevna is one of the notorious "Montenegrin duchesses," who were highly influential at the Russian court.

Presenting Militsa with his book on education, Alekseev must have kept in mind her charitable activity. From the standpoint of historical context of the Moscow "school," it is, however, interesting that a short discussion of arithmology in this book could have attracted her attention.[7]

Both Militsa and her sister Stana were ecstatically devoted to religion and mysticism (it is they who brought Rasputin to the Russian court). In particular, Militsa was actively engaged in occult sciences, and received a diploma of doctor of Hermetism in Paris.[8] Documents from Militsa's archives point to an intense relationship between Militsa, Stana, and their respective husbands, Grand Dukes Petr Nikolaevich and Nikolaĭ Nikolaevich, who exercised considerable influence on the Tsar, and French occultists.[9]

In a letter of invitation addressed to the Grand Duke Petr Nikolaevich, or to the Grand Duke Nikolaĭ Nikolaevich, the famous French occultist Sainte Yves d'Alveydre (1842–1909) promised to his addressee and the latter's wife a "true initiation," both religious and scientific: "Toute scientifique, positive et expérimentale quelle soit, elle dépasse le rêve le plus idéal de la foi Chrétienne. Elle la change en certitude absolue."[10] Then, he asks them to get acquainted before their visit with the "questionnaire of Archéomètre."[11]

This means that Militsa and her circle were familiar with some ideas of d'Alveydre's *Archéomètre* (published posthumously in 1911[12]). It concerned mystical mathematics, and "arithmologie" was one of its key words. Although its meaning had nothing to do with Bugaev's concept, the word was the same. Moreover, it was employed in connection with the attempt to "turn Christian faith into absolute certainty," which three years later Men'shikov discerned in Bugaev's theories.[13] This could actually have brought the latter's arithmology within Militsa's sphere of interests.

DOI: 10.1057/9781137338280

The fact that Alekseev presented his book to the Grand Duchess makes one wonder what kind of readership the Moscow "school" had in mind. It is reasonable to suppose that members of the "school" were not content with elaborating on a monarchal political doctrine, but tried to advertize their ideas among high-ranking and/or aristocratic readers. The tendency to expatiate on ideological matters may have been reinforced by this.

A later book of which Alekseev was the editor corroborates this supposition. During the First World War, in 1916, a volume appeared bearing the title *Nauchnaia pedagogika i russkaia shkola* [*The Scientifically Grounded Pedagogy and the Russian School*]. It mainly consisted of materials pertaining to the life of the Wilno School District, including Alekseev's circular letters.

Nauchnaia pedagogika is permeated with the Slavonic patriotism of the war times. The methods applied in the Wilno School District are presented as the realization of the pedagogical approach, originated by John Amos Comenius and culminated in the Moscow "school," on which there is a lengthy chapter.[14] Herbartianism is given here a very different place from that in Alekseev's earlier works. Now it is no more than a repetition of what had already been done by Comenius.[15]

The overall impression intended to be conveyed by Alekseev is that not only educational theories, but the whole philosophical basis of the contemporary culture was founded by the Slavs: both Leibniz, who discovered the law of continuity,[16] and the members of the Moscow "school," who explored that of discontinuity, are introduced as manifestations of the Slavonic spirit.[17]

Accordingly, Bugaev and Nekrasov are constant references. Mathematics is accorded the position of ideological foundation of the proper national system of education. In Alekseev's circular letters, arithmology emerges as an essential matter for the directors of the schools of the Wilno District to ponder over.[18]

The preface written by Alekseev is closed by the expression of gratitude to Prince Semen Semenovich Abamelek-Lazarev (1857–1916) who endowed the publication.[19] Enormously wealthy and influential, the prince was famous for his generous patronage and various scholarly pursuits. Moreover, he was a staunch Russian monarchist, and, although there is something slightly comical about Alekseev's having found a Georgian–Armenian sponsor for his Slavophil sentiments, Abamelek-Lazarev could be a valuable patron of the projects of the Moscow "school."

DOI: 10.1057/9781137338280

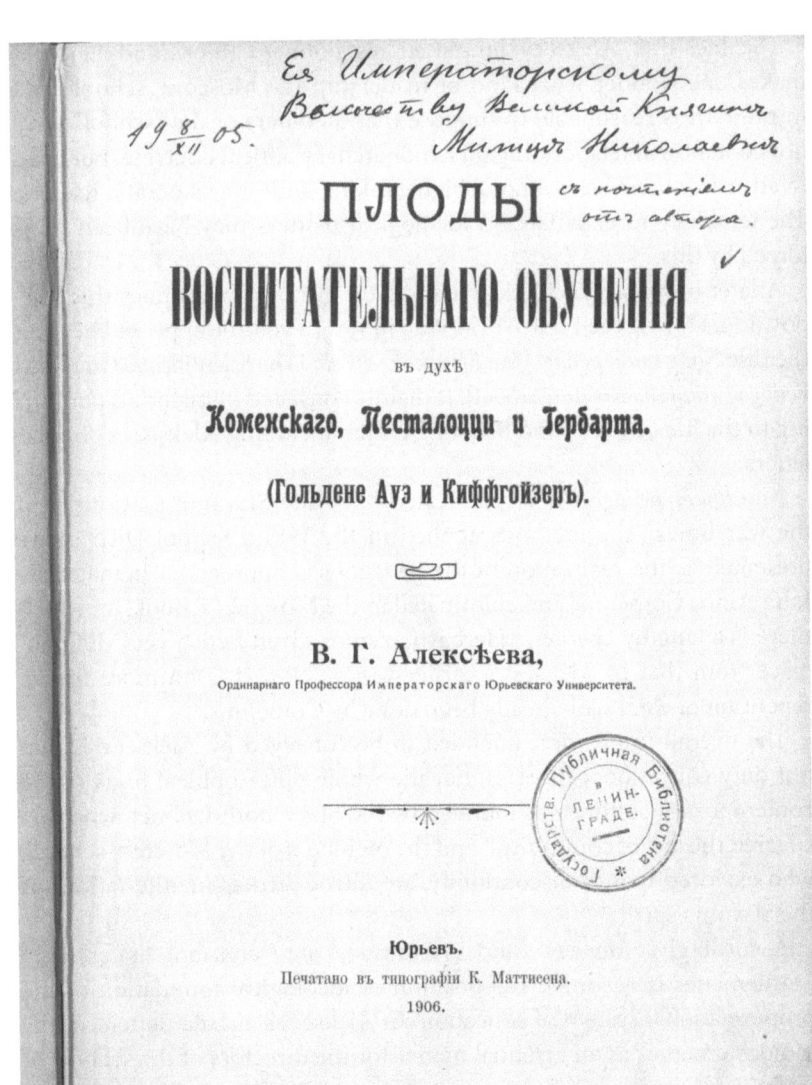

FIGURE 4.1   *The title page of V. G. Alekseev's* Plody vospitatel'nogo obucheniia v dukhe Komenskogo, Pestalotstsi i Gerbarta *(1906) presented to the Grand Duchess Militsa Nikolaevna (Rossiĭskaia natsional'naia biblioteka)*

DOI: 10.1057/9781137338280

Baron M. F. Taube

1

The second member of the Moscow "school" to be briefly considered here is especially characteristic. His name is absent from the history of mathematics and even from the history of the Moscow "school." On the other hand, he is known quite well as a representative of the so-called neo-Slavophilism, a conspicuous ideological current of the extreme Right. Baron Mikhail Ferdinandovich Taube (1855–1924) was an active member of the Black Hundred, and a celebrated author of a poem in praise of the latter[20] still to be found on anti-Semitic websites.

Taube was Nekrasov's friend and is in fact one of the most frequent references in the second edition of *Theory of Probability*. When Nekrasov writes "we," he goes on to specify: "Baron Taube and I."[21] The latter's title may have had something to do with Nekrasov's infatuation with him. At any rate, some of Taube's works are to be regarded as an integral part of the corpus of the Moscow "school."

Nekrasov and Taube wrote very favorable reviews of each other's books, those by Taube being almost as extraordinary as Nekrasov's. He was a prolific writer on various subjects. Poetry, journalism, literary criticism and philosophy were all his domains. A library formed by his writings would be almost entertaining due to their exceptional silliness,[22] had they not revealed a troubled mind deeply infected with the cannibalistic attitudes of the radical Right.

Engaged in pan-Slavism, Taube's publications were to be found in the *Slavic Century*, of which, as mentioned above, Bugaev was a reader.[23] Their political views were similar. Both shared pan-Slavic aspirations; both, which often went as a corollary to the latter, were opposed to German influences. Living in St. Petersburg, Taube started to mark his books with the imprint "Sv. [Sviato] Petrograd," long before the capital actually had its name russified. He was very keen to emphasize Slavonic origins of his philosophic and scientific authorities. Leibniz, due to his Sorbian roots, transformed in Taube's writings into the "glorious and great Slav" Lübenić or Lübnitz, the discovery of Heliocentricism into a manifestation of a Slavonic mind.[24]

Taube was a regular author of *Mirnyĭ trud*. In this sequence it is necessary to give a more precise idea of this journal, for it is in *Mirnyĭ trud* that Taube's book examining the Moscow "school" was first printed, as well

DOI: 10.1057/9781137338280

as his reviews of Nekrasov, and a latter's review of him.[25] Seen against this background the ideological message of the "school" takes on clearer contours.

2

Mirnyĭ trud was founded in Khar'kov in 1902 by Andreĭ Sergeevich Vîazigin (1867–1919), a historian, a nationalist and monarchist. Initially the journal was interesting in itself, and not only as a historical phenomenon, its program being reminiscent of the famous *Zeitschrift für Völkerpsychologie und Sprachwissenschaft* published by Moritz Lazarus and Heymann Steinthal; it was explicitly orientated towards dissemination of the ideas of A. A. Potebnîa, who developed Steinthal's ideas.[26] Accordingly, in the first issues of *Mirnyĭ trud*, there were a lot of valuable materials. If later on A. I. Sobolevskiĭ printed his articles there, it was not only the consequence of his belonging to the extreme Right, but also that of the journal's original project of publishing research in the field of linguistics, philology and history.

The ideological—nationalistic and conservative—stance of the journal was strong from the very beginning, its name referring to peaceful labor as the ground of social development in contrast to destructiveness of revolutionary turmoil.[27] The emphasis on language as the necessary basis of the worldview was intended as a remedy for "cosmopolitan dreams."[28] It is therefore not surprising to discover in one of the first issues a review of Nekrasov's *Filosofîia i logika nauki o massovykh proîavleniiakh chelovecheskoĭ deîatel'nosti*, a review which did not touch upon mathematics, but treated entirely of ideology, considering, in particular, Nekrasov's conclusions drawn from his theory of probability as a counterpart of the linguistic ideas on which the journal's program had been founded.[29]

At the time of the Russian–Japanese war and the first Russian revolution *Mirnyĭ trud* rapidly evolved into a Black Hundred organ, which entailed a dramatic change in the quality of what was published there. The journal became filled with monotonous rightist propaganda. When in 1907 Taube brought out in three installments his work *Moskovskaîa filosofsko-matematicheskaîa shkola, osnovannaîa prof. Bugaevym, i Slavîanofil'stvo Khomîakova* [*The Moscow Philosophic-Mathematical School, Founded by Prof. Bugaev, and the Slavophilism of Khomîakov*],[30] it appeared in the company of installments of A. S. Shmakov's *Evrei v istorii* [*The Jews in History*][31], those of V. F. Zalesskiĭ *Psikhicheskoe ubozhestvo iudeev* [*The Psychical Poverty of*

DOI: 10.1057/9781137338280

the Israelites],[32] and other materials of the same sort. Next year, Taube published this work as a book on the press of *Mirnyĭ trud*. To be fully aware of the implications of Slavophilism manifested in *Theory of Probability*, it is essential to realize the immediate setting of Taube's work.

3

We find in Taube explicit formulations of what Nekrasov apparently meant, but did not express clearly enough. Thus, Taube saw in Nekrasov's writings "the mathematical formula" that would bring about the downfall of "Anglo-Jewish capitalism": "Russia heeds to the voice of the motherland-loving professor; sooner or later the golden idol of Jewry and the world mercantilism will fall under the life-giving rays of the 'sun of truth' of the Enlightenment of the Orthodox East."[33]

Nekrasov's belief that arithmology provides a mathematical basis for monarchy is taken up by Taube and phrased in a clear-cut way: "the doctrine of the autocratic type of governing rests entirely on the mathematical school's view of the process of life."[34] The train of thought behind this statement is also more explicit than in Nekrasov. "The laws are firm, and the human will can change not a single <...> action, for these actions are but manifestations of <...> unchangeable laws."[35] Such is the deterministic view for which Bugaev's arithmology is a panacea, since it brings back, by way of an exact mathematical calculation, "freedom, personal originality, social independence, and autocracy of all types [samoderzhavie vsekh vidov]."[36]

Taube, as a monarchist and an ardent participant in the Black Hundred, with a pretension to polymath knowledge combined with a total lack of common sense, had every reason to be attracted by Nekrasov's conceptions. They came to be very prominent in his writings starting with his book on Bugaev. A good example is provided by a thick volume bombastically entitled *Svod osnovnykh zakonov myshleniia* ([*The Code of the Main Laws of Thinking*], 1909). Similar to Nekrasov, Taube was a distorted reflection of the contemporary fashion for Kantianism. The book could be described as pertaining to the theory of knowledge, purporting to correct Kantian philosophy with that of Khomîakov, and to support this synthesis with mathematics presented as the "only exact intellectual superlogical [sverkhlogicheskim] expression of the spiritual world."[37]

As a result, the book is thoroughly peppered with diagrams and formulas after the fashion of writers on mystical mathematics.

DOI: 10.1057/9781137338280

In fact, one of Taube's greatest concerns was occultism.[38] His *Sovremennyĭ spiritizm i mistitsizm* provides a map of the mind of a Black Hundred fighter against heretics: a mind thoroughly imbued with all sort of superstitions. At the end of the book there is a *Dictionary of foreign expressions* [*Slovar' inoiazychnykh recheniĭ*], where, among articles on vampires and larvae, the reader discovers one on arithmology.[39] The article broadly relates to Bugaev's own ideas (except for their being placed in the direct line of descent from Slavophils), but the context is suggestive of the new connotations of arithmology.

4

Through his numerical speculations Taube links the Moscow "school" with other writers of the extreme Right, who had nothing to do with mathematics, and yet touched upon what from this perspective must have seemed a cognate subject. Kabbala could have been treated either as the most typical manifestation of the Jewish mind (allegedly sterile, narrow and mendacious), or as an elevated Egyptian art stolen and degraded by the Jews.[40] In both cases having some knowledge of Kabbala was considered as a means of defense. Taking but one example, Georgiĭ Vasil'evich Butmi (1856–1917) included the most abstruse calculations in his *Vragi roda chelovecheskogo* ([*The Enemies of Mankind*], 1906).[41] The core of this popular book, written in the aftermath of the revolution, was formed by the publication of the *Protocols*.[42]

Notwithstanding Taube's references to professional mathematicians, his calculations were of the same type. Reading him we are within the same peculiar mindset haunted by the same fantasies. It is only reasonable to suppose that there was some relationship between his mathematical philosophy, nourished by Bugaev's arithmology and Nekrasov's treatment of theory of probability, and the notions of Kabbala spread among his fellows from the Right. It is not improbable, though not verifiable, that he may have entertained some notion of the mathematics of the Moscow "school" being a counterblast to the Kabbalistic practices of the Jews.[43]

5

Within the context of this book, it is particularly important to emphasize that we can find the blend of Russian Orthodoxy, monarchism, nationalism, anti-Semitism and mathematics well beyond the circle of professional mathematicians. In Taube we have a rabid conservative supporting

DOI: 10.1057/9781137338280

ПОСТРОЕНІЕ СВОДА ЗАКОНОВЪ. 139

3) Третья внутренняя троица соотвѣтственныхъ другъ другу законовъ тожества, тождества и нераздѣльной тождественности, указанныхъ буквой Б, образуетъ не треугольникъ, а точку.

4) Всѣ шесть концовъ осей упираются на окружность круга, внутри котораго помѣщается сосредоточенное скрещеніе соотвѣсныхъ осей.

Если теперь всѣ эти составныя части связать вмѣстѣ, то получится какъ-бы масонская звѣзда въ полномъ ея составѣ, съ кругомъ и крестомъ.

Сей плоскій чертежъ, безъ третьей духовной оси, — обликъ матеріалистической мысли масонства. которое по сути не духовнаго, а исключительно вещественно-душевнаго свойства.

Нашъ-же образецъ — объемень; въ немъ нѣтъ ни малѣйшей тѣни сходства съ масонскими измышленіями. Мысль соотвѣсной полноты заключается въ опредѣленіи шаровой всеобхватывающей все-

FIGURE 4.2 *A page of M. F. Taube's* Svod osnovnykh zakonov myshleniîa *(1909)*

DOI: 10.1057/9781137338280

his reactionary beliefs with mathematics. Hence the extraordinary peans made in praise of mathematics amidst writings integrally belonging to the history of the extreme Right:

> To attain scientific reliability they [freedom, personal originality, social autonomy and monarchism of all types] must be <...> put into mathematical language and numerical symbols. These thoughts ensue from the words of Bugaev himself who wants to find measure in the field of thought, will and feeling <...>. Mathematics in its limitless applicability is really powerful and wonder-working [Matematika v ee neogranichennom prilozhenii deĭstvitel'no mogushchestvenna i chudotvorna].[44]

P. A. Florenskiĭ

1

With Pavel Florenskiĭ we get to the very heart of the Russian cultural renaissance of the beginning of the twentieth century. Probably the most famous author associated with the Moscow philosophic-mathematical school, Florenskiĭ was extremely versatile and talented, until falling victim to the Soviet regime. His widespread reputation is that of a genius akin to Leonardo or Goethe.[45]

Michael Hagemeister proposed a different image of Florenskiĭ: "Man nannte Florenskij einen ‚der lichtvollsten Repräsentanten des russischen Geisteslebens'; ich halte ihn für einen Obskuranten."[46] One of the aspects of his obscurantism was his pathological Judeophobia. Thus, in his letter to Vasiliĭ Vasil'evich Rozanov (1856–1919), published in the latter's notorious *Oboniatel'noe i osiazatel'noe otnoshenie evreev k krovi* ([*Olfactory and Tactile Attitude of the Jews to Blood*], 1914) inspired by the Beĭlis affair, Florenskiĭ wrote that the only effective way of solving the Jewish question would be to castrate the Jews.[47] He added wistfully that such a solution would be incompatible with Christianity.

Florenskiĭ's writings are much too numerous and complicated to examine here in any detail. What is necessary to stress within the argument of this book is that his ideas must be regarded in closer connection to those of his teachers.[48]

His view of the universal and dominant role of mathematics, formulated when he was a student, reflected Bugaev's strong influence. Mathematicians should admit that mathematics "does not rule *de facto*":

DOI: 10.1057/9781137338280

> Her [mathematics'] power, her strength remains on paper; she possesses the documents giving her the rights to her kingdom, but she forgets that without subjects over whom she must reign, and without the land which would be submissive to her, she will only be empty air, the scheme without the schematized, the symbol without the symbolized, the right without the power of trying out this right <...>.[49]

The task is to render mathematics the real "queen of sciences."[50]

The versatility of Florenskiĭ, who wrote on an amazing range of subjects, was hardly a simple consequence of his natural predisposition. It should be compared—for all the difference between them—with Nekrasov's writings.[51] The inclination of both towards universality must have been stimulated by their intense reflection on mathematics as the "queen of sciences." The direction of Florenskiĭ's career, his switch from mathematics to priesthood and theology, should be regarded not merely as an example of the strong religious feelings characteristic of that period, but also within the context of the Moscow "school," with its striving after the new "arithmological"—both mathematical and religious—worldview, exemplified in Nekrasov's mathematical support of the Orthodox dogma of the freedom of will. Coming across Florenskiĭ's references to theosophists—for example, in his magnum opus *The Pillar and Ground of the Truth* defended as a thesis in Moscow Theological Academy—it might be useful to recall Bugaev's passion for India and the implications of this passion, both racial and mathematical, mentioned earlier in this book.

2

The last remark brings us back to a topic we will briefly address before taking leave of the members of the Moscow "school." In one of Florenskiĭ's early articles we find some more material allowing examination of their racial views. The article is entitled *O simvolakh beskonechnosti* ([*On the Symbols of Infinity*], 1904) and is devoted to Georg Cantor's conception of actual infinity.

One of Florenskiĭ's chief concerns in this work are the racial characteristics of mathematical research. Cantor's theory of actual infinity is claimed to be a natural outcome of the Jewish mind, just as the theory of potential infinity betrays its Aryan nature:

> In vague forms [v neiasnykh oblikakh] this idea of the infinity of the series of numbers emerged among the Aryans, and the thought, which had

DOI: 10.1057/9781137338280

just made that discovery, was enjoying the experimental "verification." Numerical speculations with colossal numbers in the laws of Manu, all the cosmogonical ideas, the legend about Buddha outdoing sages in counting [pobivaĭushchem v schete mudretsov], and other facts of the same kind are infused with the idea of potential infinity, though the latter, of course, is not outlined distinctly enough. This idea of bad infinity, lurking, no doubt, beyond large numbers, is meant, as it were, to crush us.[52] One would say that the idea of potential infinity is a national idea of the Aryans, chiefly of the Hindus, just as that of the actual infinity is the national idea of the Semites, mostly of the Jews.[53]

Florenskiĭ was a champion of Cantor's theory. The attempt to place it within the racial frame of reference may suggest that at that time he had had different views from those expressed in the course of the Beĭlis affair. In fact, the way that he characterizes Cantor largely derives from Vl. Solov'ev's *Evreĭstvo i khristianskiĭ vopros* ([*Jewry and the Christian Question*], 1884), an article which was profoundly philosemitic, being based on Solov'ev's firm belief in the future reconciliation of Judaism and Christianity.[54]

There is also another possibility. There may have been no abruptness in the transition from this early work on Cantor to Florenskiĭ's later racial attitudes. It was common knowledge for every intellectual at that time, who was interested in questions of race, that the Semites, who—in contrast to the Aryans—were regarded as deprived of any artistic, philosophic and scientific talents, were gifted with a powerful religious instinct, in which they surpassed the Aryans.[55] The only way to explain why a Jew made such an important discovery as that of Cantor was to represent it as a direct outcome of this religious impulse. Thus, the idea of actual infinity would have been not so much the matter of a great scientific advancement, which would be inexplicable within the frame of reference of contemporary racial theories, but that of a religious revelation.

In his later autobiography, Florenskiĭ explained his joining the Moscow Theological Academy in 1904 by his wish to study the "history of the worldview of peoples [istoriĭa mirovozzreniĭa narodov]."[56] This explanation was obviously dictated by the new Soviet surroundings, but, at the same time, it referred to Florenskiĭ's genuine interest, the article on Cantor being published just after he had entered the Academy.[57] When later on dissociated from Solov'ev's influence, this interest in national and racial character of thought (in the article on Cantor, Bobynin was another important source) acquired the intense anti-Semitic implications characteristic of the Moscow "school."

DOI: 10.1057/9781137338280

"Ableukhov" as a reflection of the Black Hundred ideology of the Moscow philosophic-mathematical school

Now let us briefly look back at the senator in Belyĭ's novel. It was by no means an accident that Belyĭ gave him the surname Ableukhov, not at all a common name. For contemporaries it could not have failed to arouse associations with a prominent member of the Black Hundred (more specifically, the Union of Michael Archangel) N. D. Obleukhov.[58] In view of all said above, it seems to be an appropriate name for a conservative senator with an emphatically foregrounded passion for geometry. It implied a reference to the political ideas of Moscow mathematicians who were thus identified, and on very good grounds, with the Black Hundred. Why this geometrical "side" of the senator was in fact a satire, despite Belyĭ's sharing some of the basic views of the Black Hundred, we shall discuss later.

The core of the senator's image may have been furnished by Bugaev, but Belyĭ's characters are often a whole mosaic of features lifted from diverse persons. In the geometrical musings of the senator concerning the ideal state, one would seem to discern the echoes of Nekrasov's unforgettable writings.[59] There is little doubt that Belyĭ read them. His father being for him a life-long influence, Belyĭ could hardly miss the most voluminous book concerning Bugaev, which was Nekrasov's *Moscow Philosophic-Mathematical School*.[60] We may assume that Belyĭ's senator, in his geometrical aspect, was a satire on the whole school of thought originated with his father, and fully expressed by his father's "most active" disciple, who came up with the idea of mathematics as a fundament for ultra-conservative politics.

Notes

1   The authors of the fullest account of Alekseev's biography put in the title of their publication 1943 as the year of his death; in the text of the publication itself they write, however, that this date remains uncertain. See V.A. Kostin, ÎU.I. Sapronov, and N.N. Udodenko, "Vissarion Grigor'evich Alekseev—zabytoe imîa v matematike (1866–1943)," *Vestnik VGU. Seriîa fizika, Matematika* 1 (2003), 139–140.
2   Kostin et al., "Vissarion Grigor'evich Alekseev," 134–135.

DOI: 10.1057/9781137338280

3   ibid., 135.

4   *Mysli k reforme nashikh shkol v dukhe pedagogiki Gerbarta*
     (Vil'na, 1881).

5   V.G. Alekseev, *Gerbart, Strumpel i ikh pedagogicheskie sistemy* (ÎUr'ev, 1907),
     3, 34; Alekseev cites this phrase also when defending Nekrasov's idea of the
     introduction of theory of probability in the school curriculum. See P.A.
     Nekrasov, ed., *Teoriia veroîatnosteî i matematika v sredneî shkole* (Petrograd,
     1915), 42. This statement is to be found in Herbart's *Pestalozzi's Idee eines ABC
     der Anschauung* (1802–1804) in Johann Friedrich Herbart, *Sämtliche Werke
     in 19 Bd.*, B. 1 (Darmstadt: Scientia Verlag Aalen, 1989 [2. Neudruck der
     Ausgabe Langensalza, 1887]), 168.

6   Cf. V. Alekseev, *K voprosu ob obrazovatel'nom znacheniii kursa teorii
     veroîatnosteî dlia sredneuchebnykh zavedeniî* (ÎUr'ev, 1914), 1–2.

7   The philosophy of the Moscow "school" is discussed in the chapter on
     Herbart, whose conception of the "real" was, according to Alekseev,
     developed and deepened by Bugaev's arithmology. See V.G. Alekseev, *Plody
     vospitatel'nogo obucheniia v dukhe Komenskogo, Pestalotstsi i Gerbarta*
     (ÎUr'ev, 1906), 19.

8   GARF, f. 669, op. 1, ed. khr. 92. The diploma delivered on September 23, 1900
     was signed by Papus.

9   There are letters from Papus, d'Alveydre, and Philippe (the notorious
     charlatan brought by Militsa to the Russian court; GARF, f. 669, op. 1, ed.
     khr. 93, 81, 82). There are also poems by d'Alveydre dedicated to Militsa (ed.
     khr. 95); see also below, nn. 10, 11.

10   GARF, f. 669, op. 1, ed. khr. 93, L. 1 – 1 ob. The letter dated September 16,
     1900 and addressed to "Monsieur" is held in an envelope marked with the
     name of the Grand Duchess Stana Nikolaevna.

11   GARF, f. 669, op. 1, ed. khr. 93, L. 1 ob. – 2. In Militsa's archives there is
     a manuscript by d'Alveydre dedicated to her and entitled *Archéomètre:
     Révélateur Homologique des Hautes Études*. It contains notes devoted to the
     meaning of letters (GARF, f. 669, op. 1, ed. khr. 96).

12   The preface of the first edition signed "Les Amis de Saint-Yves" is dated May
     23, 1911. See Saint-Yves d'Alveydre, *L'Archéomètre. Clef de toutes les religions
     et de toutes les sciences de l'antiquité. Réforme Synthétique de tous les Arts
     Contemporains* (Paris: Dorbon-Aîné, s.a), 4.

13   See Chapter 2, Section "Arithmology."

14   A.I. Shestov, and V.G. Alekseev, eds, *Nauchnaia pedagogika i russkaia shkola*
     (ÎUr'ev, 1916), 82–105.

15   ibid., 105.

16   ibid., 86 n. 1.

17   ibid., see especially 3, 105.

18   Shestov and Alekseev, *Nauchnaia pedagogika i russkaia shkola*, 113–114.

DOI: 10.1057/9781137338280

19 ibid., XVI. The materials concerning this publication are held in Abamelek-Lazarev's archives (RGADA, f. 1252, op.1, ch. 1, ed. khr. 611).

20 One of its publications is that in *Mirnyĭ trud*: M. Taube, "Chernosotenets," 5 (1906), 195. The poem is followed by Gringmut's *Guide of the Black Hundred Monarchist* (published anonymously; 196–203).

21 P.A. Nekrasov, *Teoriia veroiatnosteĭ* (S.-Peterburg, 1912), XXXVI.

22 The following example indicates the kind of mind that was fertile soil for Nekrasov's projects. It is taken from Taube's work on Pushkin's *Ruslan and Liudmila*. Taube interpreted the character of Chernomor (a dwarf with a very long beard) thus: "The name of Chernomor <…> points to the dark sea depths [as referring to the Black sea – 'Chernoe more'] and to the sphere of actions and interests of the colonial policy of the little islands of Great Britain. The little <…> hunchbacked dwarf with the 'endless moustache and the richest beard', can serve as its graphic illustration. The beard itself hints at the net of the colonies of the Anglo-Saxons, connected with metropolis by way of long voyages of the trade fleet, which entangled all the earth as if by the endless invisible grey hairs of the beard [oputavshimi vsiu zemliu tochno beskonechnymi nevidimymi sedinami borody]." See M.F. Taube, *Vseslavianskoe znachenie 'Ruslana i Liudmily'* (Sv. Petrograd, 1904), 11.

23 In the preface to his *Theory of Probability* Nekrasov cites Taube among "Bugaev's pupils and colleagues" (*Teoriia veroiatnosteĭ*, IX). There seems to be no evidence that Taube was actually acquainted with Bugaev.

24 Mikh. Ferd. Taube, *Svod osnovnykh zakonov myshleniia* (Petrograd, 1909), in particular, 133; id., "Neskol'ko slov ob A.S. Khomiakove," *MT* 10 (1905), 135. One may be surprised at somebody having Taube's surname being Slavophil and pan-Slavist, but in this he was following in a well-trodden path, both Slavophilism and pan-Slavism owing much to the German patriotic movement of the Napoleonic times.

25 M.F. Taube, "Prilozhenie osnovnykh zakonov myshleniia k voprosam statistiki i finansovogo pravomeriia," *MT* 5 (1911), 76–96; id., "Uchenie o veroiatnostiakh, kak put' k tvorcheskomu ponimaniiu dukhovnogo, dushevnogo i veshchestvennogo," *MT* 11 (1913), 147–166; P.A. Nekrasov, "Teoretiko-poznavatel'nye postroeniia v slavianofil'skom dukhe," *MT* 6–7 (1913), 280–292.

26 See A. Viazigin, "Ot redaktzii," *MT* 1 (1902), 1–2; V. Khartsiev, "Uchenie A.A. Potebni o narodnosti i natsionalizme," *MT* 2 (1902), 179.

27 Viazigin, "Ot redaktsii," 8–9. Nekrasov's calling Bugaev "a hero of peace and peaceful labor" (see Chapter 2, Section "Autocracy") is to be understood within these terms of reference.

28 Viazigin, "Ot redaktsii," 1.

29 The independent development of national cultures was allegedly corroborated by Nekrasov's treatment of probability theory as a

DOI: 10.1057/9781137338280

mathematical proof of the freedom of will. A. Kaut, "P.A. Nekrasov. Filosofiîa i logika nauki o massovykh proîavleniîakh chelovecheskoĭ deîatel'nosti (Peresmotr osnovaniĭ sotsial'noĭ fiziki Quetelet)," *MT* 4 (1902), 222–223; cf. P.A. Nekrasov, "Filosofiîa i logika nauki o massovykh proîavleniîakh chelovecheskoĭ deîatel'nosti (Peresmotr osnovaniĭ sotsial'noĭ fiziki Quetelet)," *MS* 23: 3 (1902), 585–587.

30   *MT* 10 (1907), 31–58; 11 (1907), 145–167; 12 (1907), 163–192. Cf. Taube's "Neskol'ko slov ob A.S. Khomîakove," *MT* 10 (1905), 135–136, where Taube briefly mentions Khomîakov's mathematical studies and draws the comparison of his philosophy with that of Bugaev.

31   *MT* 10 (1907), 136–161; 12 (1907), 76–122.

32   *MT* 10 (1907), 164–176; 11 (1907), 168–174.

33   Taube, "Prilozhenie osnovnykh zakonov myshleniîa," 96.

34   Taube, *Moskovskaîa filosofsko-matematicheskaîa shkola, osnovannaîa prof. Bugaevym i slavîanofil'stvo Khomîakova* (Khar'kov, 1908), 50. Cf.: "<...> the freedom of arbitrary rule [svoboda proizvola] is not only possible, but is a strict mathematical necessity of the manifestation of the creative spirit" (Taube, *Moskovskaîa filosofsko-matematicheskaîa shkola*, 7–8).

35   Taube, *Moskovskaîa filosofsko-matematicheskaîa shkola*, 8.

36   ibid., 10.

37   Taube, *Svod osnovnykh zakonov myshleniîa*, 145.

38   See Chapter 2, n. 95.

39   Taube, *Sovremennyĭ spiritizm i mistitsizm*, 218.

40   According to another version, Kabbala came from India: Blavatsky, *Isis Unveiled: A Master-Key to the Mysteries of Ancient and Modern Science and Theology, Vol. I* (Theosophy Trust, 2006), 121, n.

41   G. Butmi, *Vragi roda chelovecheskogo* 4th edn (S-Peterburg, 1907), 24–25, 122–124.

42   ibid., 43–101.

43   Cf. Chapter 5, Section "The secret radicalism of the 'school'".

44   Taube, *Moskovskaîa filosofsko-matematicheskaîa shkola*, 10.

45   Thus, the subtitle of the excellent biography of Florenskiĭ by Avril Pyman evokes "Russia's Unknown Da Vinci."

46   Quoted in Holger Kuße, "Harmonienangst. Anmerkungen zur dunklen Seite Pavel Florenskijs," in Michael Hagemeister and Torsten Metelka, eds, *Appendix 2. Materialien zu Pavel Florenskij* (Berlin u. Zepernick: Kontext, 2001) , 143. Cf. Michael Hagemeister, "Wiederverzauberung der Welt: Pavel Florenskijs Neues Mittelalter," in N. Franz, M. Hagemeister, and F. Haney, eds, *Pavel Florenskij—Tradition und Moderne. Beiträge zum Internationalen Symposium an der Universität Potsdam, 5. bis 9. April 2000* (Frankfurt am Main, Berlin, Bern, Bruxelles, New York, Oxford, Wien: Peter Lang, 2001, 41.

DOI: 10.1057/9781137338280

Cf. the following contemporary observation concerning Florenskiĭ's *Stolp i utverzhdenie istiny*: "The seminary's wisdom can be very deep as is shown by the fundamental writing *Pillar and Ground of the Truth*, but it is far from real science and is only able to establish religious truth for believers." This observation belongs to the outstanding mathematician Andreĭ Andreevich Markov (1856–1922), who was of very liberal views and in constant fight with the Moscow "school" (Oscar B. Sheĭnin, "Publikatsii A.A. Markova v gazete 'Den' za 1914 – 1915 gg." *IMI* 34 (1993), 200–201; http://www.sheynin. de/download/2_Russian%20Papers%20History.pdf, 65). "The seminary's wisdom [seminarskaîa mudrost']" is intended to be offensive, and is obviously not fair: Florenskiĭ's work is a fruit of profound education, and not of that habitually associated with seminaries. Yet, such a description is not entirely unjustified, for Florenskiĭ gives utterance to superstitions worthy of a seminarian or a Rightist like Taube. See, in particular, his deliberations on magic and politics: *Stolp i utverzhdenie istiny*, T. 1 (II) (Moskva: Pravda, 1990), 698–699 n. 268).

47 The letter was signed with "ω." Its authorship was authenticated by Michael Hagemeister ("Pavel Florenskij und der Ritualmordvorwurf," in Hagemeister and Metelka, *Appendix 2*, 62–64). This letter (in a slightly different form) was then published in the correspondence of V.V Rozanov and P.Florenskiĭ. See V.V. Rozanov, *Sobranie sochineniĭ. Literaturnye izgnanniki. Kniga vtoraîa* (Moskva – Sankt-Peterburg: Respublika; Rostok, 2010), 145–147.

48 The very fact of this connection was frequently pointed out (in particular, see S.S. Demidov, "Iz ranneĭ istorii Moskovskoĭ shkoly teorii funktsiĭ," *IMI* 30 (1986), 126–127; V.A. Shaposhnikov, "Filosofskie vzglîady N.V. Bugaeva i russkaîa kul'tura kontsa XIX–nachala XX vv." *IMI*. Vtoraîa seriîa 42: 7 (2002), 68–70), but it tends to be presented in an uncritical way. See, for example, S.S. Demidov and C.E. Ford, "On the Road to a Unified World View: Priest Pavel Florensky—Theologian, Philosopher and Scientist," in T. Koetsier, and L. Bergmans, eds, *Mathematics and the Divine: A Historical Study* (Amsterdam etc.: Elsevier, 2005), 598–599.

49 S.M. Polovinkin, "O studencheskom matematicheskom kruzhke pri moskovskom matematicheskom obshchestve v 1902–1903 gg." *IMI* 30 (1986), 150.

50 ibid., 151.

51 There was a correspondence between them. See Sheĭnin, "Publikatsii A.A. Markova," 196; http://www.sheynin.de/download/2_Russian%20Papers%20History.pdf, 63.

52 It should be noted that Florenskiĭ identifies the Aryan idea of infinity with the Hegelian concept of "die schlechte Unendlichkeit."

53 P. Florenskiĭ, "O simvolakh beskonechnosti," *Novyĭ put'* 9 (1904), 193.

DOI: 10.1057/9781137338280

54   V. S. Solov'ev, *Sochineniia v 2 tt.*, T. 1 (Moskva: Pravda, 1989), 206–256. The
      traces of Solov'ev's argument are discernible throughout Florenskiĭ's article,
      and, in particular, in his culminating comparison of Cantor's approach
      to infinity with the manner in which the Jews appeal to God. "Whereas
      others, the Aryans, recognize only the potential infinity, a 'bad' one, the
      indefinite and unlimited [neopredelennoe i neogranichennoe], the thought
      of the impossibility of the actual infinity seems to his [Cantor's] soul to
      be monstrous." In his theory Cantor manifests the same "firm insistence,"
      "almost importunity [of the Jews] in their prayers to God" (Florenskiĭ, "O
      simvolakh beskonechnosti," 233–234).

55   In particular, see on the "instinct supérieur" of the Semites as treated by
      Renan in Maurice Olender, *Les langues du Paradis. Aryens et sémites: un
      couple providential* (Paris: Seuil, 2002), 112–113, 128–129. It is also instructive
      to compare Florenskiĭ's article, written when he was about to abandon his
      mathematical career for an ecclesiastical one, with the speculations of the
      Lutheran theologian Rudolf Friedrich Grau (1835–1893), whose views, as put
      by Olender, were "opposées aux valeurs aryennes trop souvent exaltées au
      détriment de l'Église" (Olender, *Les Langues du Paradis*, 200).

56   Quoted after K.P. Florenskiĭ, "O rabotakh P.A. Florenskogo," *Trudy po
      znakovym sistemam, Vyp.* V (1971), 501.

57   When defending himself against accusations of sharing the Black Hundred
      views, Luzin described the "school of the worldview of Bugaev and Nekrasov
      P.A. [sic!]" as "a sort of utter silliness." He went on to explain: "All my
      scientific work was in sharp contradiction to their worldview. They retained
      students for graduate work at University according to their nationality, which
      I never did [Oni ostavlíali pri Universitete po natsional'nomu priznaku,
      chto u menía sovershenno otsutstvovalo]" (S.S. Demidov and B.V. Levshin
      eds, *Delo akademika Nikolaĭa Nikolaevicha Luzina* (Sankt-Peterburg: RKHGI,
      1999), 131). This statement should be compared with Luzin's reaction to
      Florenskiĭ's article on Cantor in a letter dated November 19, 1904: "I have
      read with much interest the characteristics of Cantor as a Jew. It was very
      interesting to follow the racial mathematical differences." See S.S. Demidov,
      A.N. Parshin, S.M. Polovinkin, and P.V. Florenskiĭ, eds, "Perepiska N.N.
      Luzina s P.A. Florenskim," *IMI* 31 (1989), 131. This shows that at least back
      in the 1900s Luzin's attitudes were different from those widely accepted
      nowadays. His phrase about "utter silliness" of the "school" cannot be
      construed to mean that he was entirely foreign to the ideological concerns of
      its members.

58   On him see ÎU.I. Kirîanov, *Pravye partii, 1905 – 1917. Dokumenty i materialy. V
      2 tt.*, T.1. 1905–1910 gg. (Moskva: ROSSPĖN, 1998), 12, 35, and other following
      the index. Commenting on the possible sources of the senator's name,
      N.D. Obleukhov is mentioned as a man of letters and an editor of various

DOI: 10.1057/9781137338280

reactionary organs in S.S. Grechishkin, L.K. Dolgopolov, and A.V. Lavrov, "Primechaniĭa," in Andreĭ Belyĭ, *Peterburg* (Sankt-Peterburg: Nauka, 2004), 642–643, n. 5. One of Bugaev's pupils, Leonid Kuz'mich Lakhtin (1863–1927) was probably a member of the Union of Michael Archangel (Demidov and Levshin, *Delo akademika Nikolaĭa Nikolaevicha Luzina*, 97).

59 See Appendix 1, n. 1.

60 Belyĭ was present at the sitting of the Moscow Mathematical Society in commemoration of his father where Nekrasov delivered the speech developed afterwards in the book. See "Protokol otrkytogo zasedaniĭa 16 marta 1904 g.," *MS* 25: 2 (1905), 331; cf. Lena Szilard, *Germetizm i germenevtika* (Sankt-Peterburg: Izd. Ivana Limbakha, 2002), 297; "Novaĭa matematika' i 'filosofiĭa matematiki' v *Istorii stanovleniĭa samosoznaĭushcheĭ dushi*: Aspekty aritmologii i kombinatoriki," *Russian Literature* 70: 1/2 (2011), 140.

DOI: 10.1057/9781137338280

5

The Pythagoreanism of the Moscow "School"

Abstract: *Contemporaneous comparisons of the Moscow mathematicians with the Pythagorean School, as reflected in Petersburg, contained contradictory political allusions. The Moscow "school" was considered both conservative, just as Pythagoreans had been, and radical, owing to the reputation of Pythagoras as the founder of Freemasonry.*

Svetlikova, Ilona. *The Moscow Pythagoreans: Mathematics, Mysticism, and Anti-Semitism in Russian Symbolism.* New York: Palgrave Macmillan, 2013. DOI: 10.1057/9781137338280.

DOI: 10.1057/9781137338280

Apollo

The name of the senator "Apollon Apollonovich" is also significant. It has both a connection to the mathematical context outlined above and to the topic to be discussed below, namely how and why the ideas of the Moscow "school" were satirized in *Petersburg*. This, in its turn, would give us some notion of how they were perceived at that time, which is relevant to our understanding of the later history of the Moscow Mathematical Society.

1

Belyĭ spoke of the "flight into Pythagoreanism" of his father, which needs a historical comment.[1] Being deeply interested in the history of philosophy, as well as in all kinds of esoteric knowledge, where Pythagoras was a traditional authority, Belyĭ must have known a good deal about the Pythagorean School.[2] It is therefore worth looking into some of the sources he used, or might have used, to get some general idea of how Pythagoreanism was presented in them. The following is the list of the characteristic features of Pythagoras of relevance to our topic:

1. It was still firmly believed that Pythagoras had considered everything to be a number.[3]
2. Pythagoras thought that mathematics was a necessary basis for philosophy and all knowledge. He was a great educator and had a school which assigned a leading role to mathematics.
3. Theology was one of the most important fields to which Pythagoreans applied mathematics.
4. Pythagoreans had a political doctrine based on mathematics.
5. Pythagoreans were active politicians; they were conservatives and fought against democracy.
6. Their philosophy was deeply national.

All these theses, except the last, were to be found in numerous sources.[4] The last was less frequent, but it was contained in a book well-known to Belyĭ. The notion of Pythagorean philosophy as the national philosophy of Greece occurs in *Metafizika v Drevneĭ Gretsii* ([*Metaphysics in the Ancient Greece*], 1890) by Sergeĭ Nikolaevich Trubetskoĭ (1862–1905).[5] The latter taught philosophy at Moscow University (and was its rector for a short time in 1905). Reading on the history of philosophy, Moscow mathematicians must have been familiar with his books.[6]

DOI: 10.1057/9781137338280

2

Nekrasov's vision of the Moscow philosophic-mathematical school was influenced by associations with Pythagoreanism. He evoked the Pythagorean "principle" (that is, evidently, "everything is number") as conforming to the Christian belief that the Creator arranged "all things by measure and number and weight."[7] He traced the idea of "mathematical principles of the metrical non-categorical thinking" (whatever that meant) back to Pythagoras.[8] On one instance he explicitly represented Bugaev as a follower of Pythagoras.[9] Looking upon the Moscow "school" as the revival of the "ancient principle of exact knowledge," Nekrasov envisaged using mathematics as the basis for education and every department of human endeavor. Theology was one of his most important preoccupations.

In the above mentioned *Dictionary of foreign expressions* Taube included an article *Pythagoreanism*, in which the Moscow "school" was explicitly related to the Pythagorean tradition: "In our time the metrosophic foundations of Pythagoreanism [mudromernye osnovy pifagoreĭstva] are developed by the Moscow philosophic-mathematical school, founded by prof. Bugaev <...>; its representatives are professors Alekseev and Nekrasov."[10]

It is probably not an accident that Nekrasov called the Moscow Mathematical Society a "school," or a "union," both terms being used in the studies on Pythagoreanism.[11] And even if the political involvement of Pythagoreans was not the actual reason for his deep political engagement, he could have hardly ignored this noted precedent in the history of mathematics, and it must have held some place in his imagination, the more so because the political orientation of Pythagoreans resembled that of the conservative Moscow "school."

3

In Belyĭ's imagination there was a link between Pythagoreans and politics. His definition of the Pythagorean School runs as follows: "A secret, religious-philosophical union which had political importance."[12] It is worth mentioning that one of his sources on the subject was *The Biographical History of Philosophy from Its Origin in Greece Down to the Present Day* (1845–1846) by G. H. Lewes, which, according to Belyĭ, made "everybody yawn at the times of our youth."[13] The chapter on Pythagoras, however, was far from being dull. For this chapter Lewes drew heavily on

DOI: 10.1057/9781137338280

Edward Bulwer Lytton's *Athens: Its Rise and Fall* (1843) which, whatever its historical value, was captivating. In Bulwer Lytton much space was devoted to Pythagoreans, his main interest being not their philosophy or theology, but their political engagement. Accordingly, Lewes' account of Pythagorean School accords its political activity considerable prominence, so that in the mind of Lewes' readers—and in the latter half of the nineteenth century he was a very important source for the history of philosophy in Russia[14]—Pythagoreanism must have been closely allied to politics.

It was Bugaev, with his propensity towards "Pythagorean flights," who provided the prototype for the figure of the senator. His politically active "school," fond of references to Pythagoras, was equally reflected here. This leads to the conclusion that the senator's name, however numerous its associations were, was primarily suggested by the comparison of Moscow mathematicians with Pythagoreans, whose principal God was Apollo.[15]

4

One aspect of this comparison needs to be underlined. Harmony was a recurrent term in the works of the Moscow mathematicians, who were undoubtedly aware of its roots in Pythagorean teaching. We have seen that Nekrasov referred to harmony *ad nauseam* and in particular when writing about the "truly rational state." The term became an inextricable part of the ideological discourse of the Moscow "school."

In Taube we discover a characteristic elaboration of this theme. Following Bugaev in his passion for Leibniz (that "glorious and great Slav"), Taube maintained that the very capacity of conceiving the idea of harmony was dependent upon a particular national frame of mind. A Jew was not capable of understanding harmony, because, a monist by nature, he failed to embrace the idea of trinity. A Westerner was no better, as the whole of his outlook, exemplified by Descartes, reposed on a dualism insufficient for harmony. The latter was only conceivable on the basis of Russian Orthodoxy, with its profound understanding of Trinity. Leibniz who formulated the theory of pre-established harmony was an exception, but not a surprising one, recalling his Slavic origins.[16]

Harmony was indeed a favorite word across the whole of the Moscow "school." In the minds of its members this ancient concept, along with its opposite, chaos, was invested with images suggested by contemporary

DOI: 10.1057/9781137338280

Russian politics. The very tendency to see all kinds of basic oppositions as different forms of one and the same underlying fight between the Aryans (or the Russians in this case) and the Jews was a common feature of that time.[17] What is striking here is that we detect Pythagorean allusions within the framework of the ideology of the Black Hundred. This combination found itself reflected in the name of Belyï's senator, with its first part (Apollo) referring to Pythagoreans, and the second (Ableukhov)—to the Black Hundred.[18] The result is a sort of rebus alluding to the Moscow mathematicians whose Pythagorean-like philosophy was an intrinsic part of their engagement with the extreme Right.[19]

The secret radicalism of the "school"

So far we have been concerned with the conservative side of the senator. The attempt to find its immediate frame of reference led us to the ideology of the Moscow mathematicians. Now we should examine that side of Belyï's protagonist which reflects the author's attitude to this ideology.

As has already been mentioned, Belyï's characters tend to be complex combinations of traits taken from different persons, which, among other things, was an excellent satirical device. Clashing together the incongruous features of sometimes radically opposed contemporaries hinted at their (presumably highly irritating to them) basic similarity. Thus while the "core" of the senator was formed by Bugaev and his reactionary "school," some elements of this character were inspired by a former revolutionary.

1

Physically the senator does not resemble Belyï's corpulent father. At the same time his appearance is unmistakably suggestive of Lev Tikhomirov as portrayed in Belyï's memoirs: a little man with quick movements and light step, enclosed in a "cube" of his room (like that of the senator's cubical carriage), reminding one of a bat,[20] or else of an Egyptian mummy.[21] All these traits are also those of Ableukhov, which makes us take a closer look at their original owner.

Lev Aleksandrovich Tikhomirov (1852–1923) started out as a revolutionary. He was one of the leaders of "People's Will," the clandestine organization already mentioned. After the assassination of Alexander II, he had to flee abroad. Several years later, after a spiritual crisis resulting

DOI: 10.1057/9781137338280

in the wholesale retraction of his political ideas, Tikhomirov appealed to the Tsar Alexander III for clemency and was allowed to return to Russia. Eventually, he became one of the most prominent conservative ideologists, the editor of the conservative newspaper *Moscow Gazette* and the author of the fundamental treatise *Monarkhicheskaîa Gosudarstvennost'* (*[Monarchical Statehood]*, 1905). Among his conservative adversaries, however, he had the reputation of a secret revolutionary with ideas dangerous for the state.[22]

It might well have been that Belyĭ got to know Tikhomirov through his father. He does not mention this in his memoirs, from which he had to remove all traces of his father's connections with conservatives. Nevertheless, his father, acquainted with the former editor of the *Moscow Gazette* (the above-mentioned V. A. Gringmut), is likely to have been acquainted with its new editor. Moreover, it is known that Bugaev was interested in the *Apocalypse*, which was a possible point of contact with Tikhomirov, a great expert on it. It is precisely in this capacity that Tikhomirov first appears in Belyĭ's memoirs.[23]

It is not, however, the conservative side of Tikhomirov, which may have brought him within the circle of Bugaev's acquaintances, that Belyĭ alludes to in *Petersburg*. The portrait in Belyĭ's memoirs reflects Tikhomirov's reputation of being a secret revolutionary. In the novel the passage describing the senator's love for geometry and his geometrical meditations in the "center of the black, perfect and the satin-lined cube" concludes with the observation that he was born for solitary confinement. This is paralleled by the passage in Belyĭ's memoirs where Tikhomirov, in his (and the senator's) bat-like manner dash around the "cube" of his apartment like a "captive."[24] This suggests that the "solitary confinement" for which the senator was born is not just a pun-like development of his love for cubes, but a trait borrowed from a former revolutionary. From the dream of the Ableukhov the younger, we learn that his father's conservatism was aimed at the destruction of the "Aryan world."

2

Now one begins to grasp the full significance of the geometry associated with the senator. Not only Ableukhov's conservatism is reflected in his geometrical meditations but his secret radicalism as well.

In fact, if the conservative connotations of geometry through its metonymical link with Moscow mathematicians need reconstructing, its radical meanings are much more easily perceived. Bearing in mind

DOI: 10.1057/9781137338280

the radical component of the character of the senator, we notice that his passion for geometry hints at a common characteristic of the mind of a revolutionary.

In this connection, one should mention a work that gained wide popularity in Russia after the revolution of 1905, *Les origines de la France contemporaine* (1876–1894) by Hippolyte Taine. A part of this work was published by "Musaget,"[25] and Belyĭ is very likely to have actually read it.[26] Taine provides us with a good example of the motif of geometry used in the political satire of radicals. This is how he describes a Jacobin:

> Son principe est un axiome de géométrie politique qui porte en soi sa propre preuve; car, comme les axiomes de la géométrie ordinaire, il est formé par la combinaison de quelques idées simples, et son évidence s'impose du premier coup à tout esprit qui pense ensemble les deux termes dont il est l'assemblage. L'homme en général, les droits de l'homme, le contrat social, la liberté, l'égalité, la raison, la nature, le peuple, les tyrans, voilà ces notions élémentaires: précises on non, elles remplissent le cerveau du nouveau sectaire; souvent elles n'y sont que des mots grandioses et vagues; mais · il n'importe. Dès qu'elles se sont assemblées en lui, elles deviennent pour lui un axiome qu'il applique à l'instant, tout entier, en toute occasion et à outrance. Des hommes réels, nul souci: il ne les voit pas; il n'a pas besoin de les voir; les yeux clos, il impose son moule à la matière humaine qu'il pétrit; jamais il ne songe à se figurer d'avance cette matière multiple, ondoyante et complexe, des paysans, des artisans, des bourgeois, des curés, des nobles contemporains, à leur charrue, dans leur garni, à leur bureau, dans leur presbytère, dans leur hôtel, avec leurs croyances invétérées, leurs inclinations persistantes, leur volontés effectives. Rien de tout cela ne peut entrer ni se loger dans son esprit; les avenues en sont bouchées par le principe abstrait qui s'y étale et prend pour lui seul toute la place. Si, par le canal des oreilles ou des yeux, l'expérience présente y enfonce de force quelque vérité importune, elle n'y peut subsister; toute criante et saignante qu'elle soit, il l'expulse; au besoin, il la tord et l'étrangle, à titre de calomniatrice, parce qu'elle dément un principe indiscutable et vrai par soi.—Manifestement, un pareil esprit n'est pas sain: des deux facultés qui devraient tirer également et ensemble, l'une est atrophiée l'autre hypertrophiée; le contrepoids de faits manque pour balancer le poid des formules. Tout chargé d'un côté et tout vide de l'autre, il verse violemment du côté où il penche, et telle est bien incurable infirmité de l'esprit jacobin.[27]

Ableukhov's journey at the beginning of the novel resembles this portrait. He is moving enclosed in the "black cube" of his carriage. He carefully avoids looking at what is around. The streams of people annoy him as

DOI: 10.1057/9781137338280

"matière multiple, ondoyante et complexe" would a Jacobin. The mind of the senator is virtually filled with geometry, and his love for the straight line recalls "procédé rectiligne" which is that of a Jacobin.[28]

We cannot enter here into the history of the motif of geometry in the conservative criticism of radical politics.[29] It is sufficient to indicate that Belyĭ must have used it intentionally.[30] Whereas the association between geometry and revolution is now mostly known through French art at the time of the French revolution, as well as the Russian avant-garde, that is, as a part of revolutionary discourse, at the date with which our study is concerned this association was equally used for the purpose of criticizing political radicalism.

Nekrasov's theories were explicitly criticized as potentially subversive. In 1905 there came out an article bearing a characteristic title *Filosofiia "moskovskoĭ filosofsko-matematicheskoĭ shkoly i ee otnoshenie k intellektualizmu filosofov XVIII i èkonosmicheskomu materializmu K. Marksa* [*The Philosophy of the "Moscow Philosophic-Mathematical School" and Its Relationship to the Intellectualism of the Philosophers of the Eighteenth Century and to the Economical Materialism of K. Marx*].[31] Nekrasov's projects are regarded here as fundamentally belonging, by the very nature of their utopian character, to the tradition of radical thought. It is stated that the attempts to realize such projects more often than not "rapidly and inevitably lead to revolution."[32] The senator, engrossed in geometrical–political schemes, and committed to both conservatism and destruction, was quite natural in this context, and probably owed something to contemporary reactions to the philosophy of the "school," exemplified in the critique quoted.

3

It was easy subsequently for Belyĭ to present his father's arithmology in such a light that made it cognate to revolutionary thought.

Belyĭ set out to write his memoirs in 1928, when Bugaev was already remembered mainly as a reactionary.[33] Moreover, being a reader of the main Soviet philosophical journal *Pod znamenem marksizma* [*Under the Banner of Marxism*], Belyĭ was aware of the fact that mathematics had become an object of Marxist critique.[34] This influenced his way of writing about his father. In Belyĭ's memoirs Bugaev is rebellious and unconventional; he is profoundly opposed to the retrograde professors of Moscow University. Arithmology is represented as an organic part of

DOI: 10.1057/9781137338280

this portrait: "As a mathematician, he included in the theory of evolution the revolutionary role of the leap, break, probability, quality."[35] This suggests that Bugaev's emphasis on discontinuity agreed with the idea of the social revolution. If we look once more into Bugaev's *Mathematics and the Scientific-Philosophical Conception of the World*, the word "revolutionary" occurs in his discussion of the positive outcomes of the analytical conception of the world:

> L'idée que le développement social s'accomplit au moyen d'un progrès lent et continu de tous les éléments de la societé, cette idée s'affirme de plus en plus. Dans les considérations historiques contemporaines les théories évolutionistes prennent le dessus sur les théories révolutionnaires. La science a commencé à remplacer la doctrine. Les doctrinaires ont peu à peu commencé à céder le pas aux vrais savants.[36]

The concentration on discontinuity in Bugaev's case was however by no means associated with advocacy of revolutionary changes. Having cited the same passage in one of the severest attacks on the Moscow "school" published the same year as the first volume of Belyĭ's memoirs, the Marxist physicist Vasiliĭ Petrovich Egorshin (1898–1985) summed it up as follows: "In a word, mathematics serves a firm bulwark for conservative statesmanhood in their fight with revolution <...>."[37]

The way arithmology was used by Bugaev's followers as a scientific buttress for autocratic rule was in a sense closer to Bugaev (who probably would not have endorsed it) than the interpretation of arithmology as conforming to Soviet ideology.

Yet, this transformation was an easy one, not only due to the pressure of the new regime, but also due to certain ideas widely circulated in Belyĭ's times. Thus, in anarchism, which he had studied when young, the refutation of continuity came to be regarded as a corroboration of a revolutionary doctrine.[38] Still more pertinent for the context of his memoirs was the Hegelian conception of qualitative leaps, which occur when quantitative changes pass into qualitative ones. Born within the same framework of reaction against *lex continuitatis* as arithmology, this conception was taken up by Marxism and, in particular, applied to describe the mechanism of social revolutions. The emphasis on "leaps" as opposed to "imperceptible transitions" was a regular part of Marxist discourse,[39] which explains Belyĭ's hope to represent arithmology, once employed as an ideological tool by the ultra-reactionary Moscow "school," as acceptable in the eyes of the revolutionary state.[40]

DOI: 10.1057/9781137338280

4

One further relevant point is that Nekrasov ended up by embracing Soviet rule. His obituary, signed "S. Uritskiĭ," was published in one of the most influential Soviet newspapers *Izvestiîa* [*News*].⁴¹ The very fact of this publication testifies to Nekrasov's efforts to establish connections with the new authorities.

Still more remarkable is the family name of the author of the obituary, who, as one learns from the text, was Nekrasov's acquaintance. Six years earlier, *Izvestiîa* printed an obituary of Moiseĭ Solomonovich Uritskiĭ (1873–1918), the sinister "Torquemada of the Communist Manifest," who was assassinated by the poet Leonid Kannegiesser (1896–1918). The central square of St. Petersburg (Petrograd at that time), Dvortsovaîa [Palace] square, where the assassination took place, was to bear the name of Uritskiĭ for many years onwards. His nephew, Semen Petrovich Uritskiĭ (1895–1938), became a prominent Soviet military commander, and it is probable that he was the author of the obituary. Another possibility is the latter's cousin, Semen Borisovich Uritskiĭ (1893–1940), the main editor of the newspaper *Gudok* [*Factory Whistle*] to be later appointed director of the All-Union Book Chamber. Anyway, the author is very likely to have been a relative of the once immensely powerful and terrible Chairman of Petrograd Cheka. Evidently, Nekrasov had been looking for influential protectors. That the Uritskiĭs were Jews is also curious and not without relevance in this context.

Although it is rather amazing to read an obituary of such a devoted monarchist as Nekrasov had been, signed by a name intimately associated with frightful atrocities of the Soviet regime, there is nothing surprising about it in view of Nekrasov's obvious belonging to the "cujus regio, ejus religio" school, with which his stance on mathematics as a means of defending the "highest truth" against the "fatal laws" accorded perfectly.

The last detail to mention in connection with this obituary is a note to which Uritskiĭ refers as having been written by Nekrasov, just before going into hospital. According to the former, the latter asked him therein "to participate in the formation of a scientific Marxist circle [uchenogo marksistskogo kruzhka] for the study and use of his [Nekrasov's] works." In the preceding paragraph Uritskiĭ speaks of certain "mathematician-communists" setting out to purify Nekrasov's writings from metaphysics in order to demonstrate "all their value."⁴² Although it is not clear, whether

DOI: 10.1057/9781137338280

the activity of those "mathematician–communists" was the result of Nekrasov's request, or had preceded it, it would seem that Nekrasov's zeal to please the new regime may have contributed to the subsequent campaign of persecution of the Moscow Mathematical Society. Whoever those "mathematician–communists" were, they were bound to become close students of Nekrasov's works, and could not fail to see the character of his ideological message. One is tempted to quote (not for the first time in the course of the present study) Nabokov's words about "people of limited intelligence with a great talent for mathematics."[43]

5

It is not entirely improbable that Nekrasov was able to develop a sort of blindness to the meaning of his pre-Soviet books. There was a certain similarity between his ideological projects, hostile to capitalism and centered on rationality, and the communist ones. Nekrasov's political speculations were inspired by Plato's ideal state, just as those of many radicals were.

There was not much difficulty in rewriting Plato for the ends of rightist propaganda. The rationality that Nekrasov alluded to in his ideological writings is first and foremost divine rationality. Mathematics is perceived as a tool of penetrating into divine wisdom and thus of obtaining necessary clues as to how Autocracy, the earthly reflection of the divine order, must function.

Emphasizing the Pythagorean harmony of the state, Nekrasov omitted those traits of Plato's "fair city" which were elaborated by utopists and radical writers. Pythagorean "communism," the famous "κοινὰ τὰ τῶν φίλων" principle ("the possessions of friends are common"), had no place in Nekrasov's "truly rational state." Nevertheless, this omission was not enough to make his speculations essentially dissimilar from the tradition of radical thought following in the trail of the *Republic*. Alluding to the Moscow mathematicians, Belyï's Apollon Apollonovich is equally reminiscent of the Sun-Metaphysic ruling over Campanella's city with its Pythagorean rites.[44]

6

For Belyï, geometry was not merely the sign of a revolutionary whose activities were destructive because his mind was too abstract. Nor was the radical "aspect" of this motif, simply an echo of some contemporary

DOI: 10.1057/9781137338280

polemics with the Moscow "school," which could have been re-enforced by pointing to the above mentioned convergence of arithmology with anarchism. For somebody who, like Belyĭ, was well-versed in the rightist critique of radicalism, geometry was closely bound up with Freemasonry. Meditations on geometrical figures performed by a secret destructor of the state could not be free from such connotations. Not only was Belyĭ curious about the Masons, but he was also frightened by them.[45]

The "geometrical" state that the senator dreams about is entirely rational; it is homogenous and implies equality (every citizen will have his "square"). It covers the whole universe.[46] It thus responds to what one knew about Masonic political aspirations. "*Reason* will be the only legislator," reported a horrified author of *Mirnyĭ trud* in an article written after the revolution of 1905 about the secret policy of Freemasonry. Monarchies and nationalities "will disappear from the face of the world";[47] there will be but one world state or a world tyranny, which has already started to form, since the government—this point had gained strength by the year 1911—has been overtaken by the Jews and Masons.[48] Combined with the fact that the Ableukhovs are partly Semitic, the senator's taste for geometry is to be related to this context. The "serpent coils" which emerge in the senator's meditation ("the entire spherical surface of the planet should be embraced, as in serpent coils, by blackish gray cubes of houses") must refer to the already mentioned "symbolic snake."

An avid reader of occult literature, Belyĭ certainly knew that there was a tradition of crediting Pythagoras with the foundation of Freemasonry[49] so that the latter was sometimes described as a revival of Pythagoreanism.[50] The name "Apollo" hinting at the Moscow mathematicians with their Pythagorean outlook suggests its Masonic character. The mathematicians are exposed as unconscious allies of Masons. The senator's love of geometry reveals both his overt conservatism and his secret commitment to the Masonic cause bringing about destruction and chaos.

7

In fact, Nekrasov's writings, and, in particular, the *Moscow Philosophic-Mathematical School* gave some reason—at a time of acute awareness of alleged pervasiveness of Masonic influence—of suspecting the author of sympathizing with Masonic ideas. He was obsessed with the metaphors of "building [stroitel'stvo]" and "edifice [zdanie]." He referred to

DOI: 10.1057/9781137338280

"architects" and "masons" who work over the building of the cultural state.[51] Combined with his "Pythagorean" vocabulary (harmony, mathematical and geometrical metaphors, the soul of the world[52] etc.), which was widely used in Masonic literature, the resemblance strikes one as curious. Nekrasov wrote of Freemasonry as a "secret state [taĭnoe gosudarstvo]" belonging to the "kingdom of falsehood [tsarstvo lzhi],"[53] or else as "contemptible allies of the kingdom of falsehood and chaos."[54] Yet, his vocabulary probably testifies to a more elaborate attitude. Another member of the "school," Taube, was anxious to distinguish his own use of geometry, apparently in the same vein of mystical arithmetic as that which was practiced by Masons, from theirs.[55] On the other hand, he envisaged an organization that could become an alternative to, and a defense against, Masonry. His idea was a union, loyal to Russian Sovereign and Orthodoxy, which would use the symbolism of carpenters [drevodely].[56] Trying to explain Nekrasov's predilection for metaphors that had Masonic connotations, one wonders whether the Moscow "school," associated in his mind with Pythagoreanism, was not intended to be the true Pythagorean revival in contrast to the false Masonic one.

8

Finally, in Belyĭ's archives there is an important document *Bibliografiĭa po filosofii* [*A Bibliography of Philosophy*] apparently related to his unfinished *Istoriĭa stanovleniĭa samosoznaĭushcheĭ dushi* [*A History of the Evolution of the Self-Knowing Spirit*] conceived as a grandiose canvas of the history of thought "over the last five centuries."[57] In this bibliography covering more than 90 pages, there is a part on Greek philosophy that includes a section bearing the title "O Pifagore—mificheskom [On Mythical Pythagoras]." Among other works (13 in number), three are of particular interest here: August Gladisch, *Die Pythagoreer und die Schinesen* (1841); Leopold von Schroeder, *Pythagoras und die Inder* (1884); Édouard Schuré, *Les Grands Initiés* (1889).[58] The expression "mythical Pythagoras" does not imply distrust in these sources, for next to them Belyĭ cited such works as Eduard Zeller's *Grundriss der Geschichte der Griechischen Philosophie* (1883) or A. V. Vasil'ev's *Integer* (1919)[59], which he undoubtedly considered to be reliable.

The full title of the first of the mentioned works is *Einleitung in das Verständniss in der Weltgeschichte. Erste Abteilung: Die alten Schinesen und die Pythagoreer*. August Gladisch (1804–1879), who was a pupil of famous sinologist Jean-Pierre Abel-Rémusat (1788–1832), undertook to develop the latter's observation on the similarities between certain doctrines

DOI: 10.1057/9781137338280

of the Chinese philosophy, on the one hand, and Pythagoreanism and Platonism on the other.[60] Gladisch called Pythagoreans "die Hellenischen Schinesen," tracing their system directly back to that of the Chinese, who held (as he believed) the same moral, numerical, and musical philosophy. Anticipating critical remarks about the absence of any channel through which such influence could be transmitted, Gladisch came up with an extraordinary reply. He supposed that the Greeks had actually communicated with the Chinese, but the Greek name for the latter was the Hyperboreans. According to Gladisch, there had once been the Chinese in the North (in Scythia, "das heutige Rußland"); they had worshiped Apollo, and in their "musical worldview and musical morals [in ihrer musikalischen Weltansicht und musikalischen Sittlichkeit]" had surpassed all the peoples of ancient times. This explains why the tradition connects Pythagoras with the Hyperboreans.[61]

Chinese connotations of Pythagoreanism were to become of particular relevance for Belyĭ in his novel *The Christened Chinaman*, with its protagonist Professor Letaev resembling a Scythian and advocating Pythagorean ideas. More importantly for *Petersburg*, the appearance of Gladisch' study in *Bibliografiia* corresponds to Belyĭ's general interest towards alleged Eastern origins of Western thought, an interest that is equally manifest in the presence of Schroeder and Schuré in the bibliography, who both depicted Pythagoreans as heirs of Indian wisdom. Also of relevance for the context of *Petersburg* is that Belyĭ may well have discussed this subject with Medtner, whose notes on the above mentioned study on Chamberlain include the following:

> Now it is proved that Greek philosophy was only under the influence of Indian philosophy, and not at all under that of Egyptian or, in general, Semitic thinking.
>
> Pythagoras is a disciple of the Indians. The Pythagorean path of the Hellenic philosophy was right (for example, astrophysics); but Aristotle and neo-Platonists brought confusion and retarded the Aryan culture for a long time.[62]

Encountering here the observation that Pythagoreans had already discovered what later was rediscovered by Copernicus,[63] it is reasonable to suppose that the Pythagoreanism of Ableukhov the elder, and the Kantianism of Ableukhov the younger, among other things, served to stress their similarity. The Kantian meditations of Nikolaĭ Apollonovich include a parodic reference to the famous comparison between Kant and Copernicus.[64] Through Copernicus, Kant was regarded as an offspring of the ancient Pythagorean teaching, or all of them—Pythagoras, Copernicus and Kant—as representatives of the same line of thought, which for Medtner or Chamberlain embodied the spirit of the Aryans.

The Kantian meditations of Ableukhov the younger exhibit his belief in the Aryan character of Kantianism, which in his dream (and Belyĭ's articles of that time) is revealed as Semitic, and all the more Semitic and dangerous for that it pretends to be Aryan.[65] A similar train of thought must have been linked to the Pythagoreanism of Ableuchov the elder: his allegedly Aryan Pythagoreanism was a particularly dangerous manifestation of his Semitic mind.[66] At the time of writing *Petersburg*, Belyĭ's quarrels with Medtner had already begun, so that the "Pythagorean" line of the novel perhaps was connoted by a subtle parody of the former's beliefs.

The Ableukhov coat of arms

1

At the end of the Chapter the First, the reader is told that the senator in his black carriage will forever pursue him. On this carriage one sees the senator's coat of arms depicting "a unicorn goring a knight,"—an image which haunts not only Ableukhov the younger, desperately trying to find who "gores" him, but the reader as well. The Ableukhov coat of arms has been often discussed, but its meaning has never been sufficiently clarified.

The idea of a unicorn attacking a knight was prompted by Conan Doyle's story *Playing with Fire* (1900).[67] In this story an occultist undertakes to prove the theosophical tenet that thoughts are as material as things and that imagining a thing amounts to making it ("That is why an evil thought is also a danger"). During a séance he succeeds in illustrating his point by materializing a unicorn which nearly kills everybody present.

Apollon Apollonovich, as stated in the beginning and reiterated throughout the novel, is characterized by a peculiar trait: "his cranium was becoming the womb of thought-images, which at once became incarnate in this spectral world."[68] All the characters and events of *Petersburg* originate in the thought of the senator. Ableukhov's idea of a

DOI: 10.1057/9781137338280

terrorist gives birth to the terrorist Dudkin, preoccupied with the idea to kill the senator. The dangers which threaten the Ableukhovs are of the same nature as Conan Doyle's unicorn, hence the latter on their coat of arms.[69]

Playing with fire was a crude elaboration of those theosophical ideas which attracted Belyĭ.[70] Yet the respective passages in the novel are a brilliant parody of the whole theosophical argument concerning the power of thought. It is difficult to define the nature of this parody. I would like to call attention only to a small, but relevant point. Belyĭ did not read English. Most probably, he read Conan Doyle's story in the translation of the mentioned Nikolaĭ Obleukhov.[71] This name could therefore have been associated in Belyĭ's mind not only with the Black Hundred, but with some degraded form of theosophy. Writing an autobiographical novel on a period of life that was over, and now considered as profoundly erroneous, Nikolaĭ Ableukhov was an appropriate name for a self-parody, as it came to be linked with two themes of primary importance for Belyĭ—that of nationalism combined with racism, and that of the creative power of thought, which Belyĭ had elaborated in a strange framework of Kantianism fused with theosophy. Obleukhov stood for the Black Hundred, whose appalling vulgarity must have been perceived as a distorted vision of Belyĭ's political ideas, and—through Obleukhov's translation of Conan Doyle—for theosophical banalities, which caricatured Belyĭ's mystical and philosophical quest.

This is only a conjecture, useful to highlight certain facts in order to establish the connections between them. We are on safer ground when guided by traces of Conan Doyle's unicorn in the novel. Conceived by the senator's thought (and hardly by accident a reader of Conan Doyle), the terrorist Dudkin inherits the senator's capacity of giving birth to things he thinks about. The one that nearly kills him is the state, which, after the manner of Zarathustra, he imagines in the shape of an idol, a statue, which in this case is the Bronze Horse. The latter visits Dudkin in his attic. For all the difference between Conan Doyle's unsophisticated prose and the remarkable strength of Belyĭ's, some details of the description of the attack of the unicorn are reproduced in the latter's representation of the visit of the "Copper guest," a close variation on the attacking "white horse" from *Playing with fire*.

It is not only the capacity for engendering mental monsters that Dudkin inherits from the senator. Looking at him as a continuation of Apollon Apollonovich, both menacing the latter and taking deeply after

DOI: 10.1057/9781137338280

him, one begins to comprehend the meaning of one of the culminating moments of the novel, when, after foretelling the coming apocalypse, Dudkin appeals to the Sun:

> <...> the final Sun will rise in radiance over my native land. Oh Sun, if you do not rise, then, oh, Sun, the shores of Europe will sink beneath the heavy Mongol heel, and foam will curl over those shores. Earthborn creatures once more will sink to the depths of the oceans, into chaos, primordial and long forgotten.
>
> Arise, oh Sun![72]

The sun as a god capable of averting the coming chaos, personified by the barbaric hordes, reminds one that Dudkin is the senator's brain-child.[73] The appeal to the Sun, which strikes one as an oddity, falls into place if seen as an echo of the Pythagorean worship of the sun.[74]

2

The argument has come full circle and the opposing ideologies turn out to be basically the same. Belyï's preoccupation with "circular motion" in philosophy and ideology,[75] reflected in the character of Ableukhov, brings to mind the later history of the Moscow Mathematical Society. The main prototype of the senator was a conservative who had established a school of mathematicians that aspired to make of mathematics a highly ideological state discipline. In Soviet times this school was condemned not because of the attempt to attach mathematics to ideology, but because ideology now served a different state, while mathematics continued to be regarded by the new ideologues as inseparable from ideology. The fight (or that part of the fight which concerned the Moscow philosophic-mathematical school) was conducted on the same ground, formed by the same fallacious view of the nature of science.

To ignore the elements of truth about the Moscow "school" in the writings of their Soviet critics, however grotesque these writings are, and to form our notion about the school by way of simple reversal of their Soviet opponents, would condemn us to move in a historical counterpart of the vicious circle just described. The "little senator," some of whose essential traits were suggested by intense meditations over "circular motion" in thought, can help break this vicious circle of automatic value judgments and offer a different view of the Moscow philosophic-mathematical school.

DOI: 10.1057/9781137338280

Notes

1   See Chapter 2, Section "Arithmology." Pythagoreanism is evoked in some studies on the Moscow "school" but only vaguely. See the preface by S.S. Demidov, S.M. Polovinkin, and P.V. Florenskiĭ to: P.A. Florenskiĭ, "Chernovik vystupleniia na otkrytii studencheskogo matematicheskogo kruzhka pri Moskovskom matematicheskom obshchestve," *IMI* 32–33 (1990), 468; S.M. Polovinkin, "Moskovskaia filosofsko-matematicheskaia shkola (Obzor)," *Referativnyĭ zhurnal. Obshchestvennye nauki v SSSR. Seriia 3. Filosofiia* 2 (1991), 62; M.A. Prasolov, "'Tsifra poluchaet osobuiu silu' (Sotsial'naia utopiia Moskovskoĭ filosofsko-matematicheskoĭ shkoly)." *Zhurnal sotsiologii i sotsial'noĭ antropologii* 10: 1 (2007), 45–46; Lena Szilard, *Germetizm i germenevtika* (Sankt-Peterburg: Izd. Ivana Limbakha, 2002), 298–299. A valuable discussion of Florenskiĭ's interpretation of Pythagoreanism is to be found in Anke Niederbudde, *Mathematische Konzeptionen in der russischen Moderne: Florenskij—Chlebnikov—Charms* (München: Verlag Otto Sagner, 2006), in particular, 25. I attempted to tackle this problem in the above cited article "Moskovskie pifagoreĭtsy," in S.N. Zenkin, ed., *Intellektual'nyĭ iazyk epokhi: Istoriia ideĭ, istoriia slov* (Moskva: Novoe literaturnoe obozrenie, 2011), 117–141.

2   See, in particular, Andreĭ Belyĭ, *Simvolizm* (Moskva: Musaget, 1910), 126, 516 n. 5, 546 n. 5; Belyĭ's notes on Pythagoreanism which were most probably taken around the time of his work on *Symbolism* are held in RGB (f. 25, k. 31, ed. khr. 19, L. 19).

3   The convincing argument against this belief is to be found in L. Zhmud, *Wissenschaft, Philosophie und Religion im frühen Pythagoreismus* (Berlin: Akad. Verl., 1997), 261–279.

4   From those in Bugaev's library see, for example, the cited works by Moritz Cantor and Herman Hankel.

5   Sergeĭ Nikolaevich Trubetskoĭ, *Metafizika v Drevneĭ Gretsii* (Moskva: Mysl', 2003), 216; Andreĭ Belyĭ and Ivanov-Razumnik, *Perepiska* (Sankt-Peterburg: Atheneum; Feniks, 1998), 435, 442 n. 82.

6   See Chapter 2, n. 31.

7   P.A. Nekrasov, "Moskovskaia filosofsko-matematicheskaia shkola i ee osnovateli," *MS* 25: 1 (1904), 6.

8   P.A. Nekrasov, "Logika mudrykh liudeĭ i moral' (Otvet V.A. Gol'tsevu)," *VFiP* 70: 5 (1903), 902.

9   P.A. Nekrasov, *Teoriia veroiatnosteĭ* (S.-Peterburg, 1912), XV. Cf. P.A. Florenskiĭ, "Pifagorovy chisla," *Trudy po znakovym sistemam*, V (1971), 504–506.

10   Mikh. Ferd. Taube, *Sovremennyĭ spiritizm i mistitsizm* (Petrograd, 1909), 229. The comparison of Moscow "school" with Pythagoreans was drawn not

DOI: 10.1057/9781137338280

only by its members. The mathematician Dmitriĭ Dmitrievich Mordukhaĭ-Boltovskoĭ (1876–1952), who criticized their philosophy, called the "school" the "revived Pythagoreanism [vozrodivshimsia pifagoreĭstvom]": "O zakone nepreryvnosti," *VFiP* 87: 2 (1907), 174; cf. M. Men'shikov, "Zvezdy i chisla," *Novoe vremia*, № 9990, December 25 (January 7), 1903, 7.

11 Whereas "school" sounds natural as a synonym for "society," the Russian "union" is rather bizarre in this context. Speaking of Pythagoreans, Belyĭ also employs this word (*Simvolizm*, 546 n. 5).

12 RGB, f. 25, k. 31, ed. khr. 19, L. 19.

13 Belyĭ and Ivanov-Razumnik, *Perepiska*, 389.

14 Bugaev had Lewes' *History* in his library (ORK i R NB MGU, f. 41, op.1, ed. khr. 252, L.72 ob.) Evoking Pythagoras in his *Mathematics as a Scientific and Pedagogic Instrument*, N.V. Bugaev referred to Lewes (*Matematika kak orudie nauchnoe i pedagogicheskoe* (Moskva, 1869), 18–19.

15 It is probably not irrelevant that Pythagoras assuming the powers of a senator turns up in Bulwer Lytton. See Edward Bulwer Lytton, *Athens: Its Rise and Fall*, Vol. II (Leipzig: Bernard Tauchnitz, 1843), 239; cf. Lewes, *Istoriia filosofii ot nachala ee v Gretsii do nastoiashchikh vremen* (S.-Peterburg, 1865), 21. One of the preliminary versions of the title of the novel, "Laquered carriage [Lakirovannaia kareta]," is perhaps another sign of the link between Apollon Apollonovich and Pythagoras. The principal character of the story *A Carriage* [Koliaska] by Belyj's favorite, Gogol, was called Pifagor Pifagorovich.

16 Mikh. Ferd. Taube, *Svod osnovnykh zakonov myshleniia* (Petrograd, 1909), 29, 133. Cf. on Nekrasov's conception of epistemological "harmony" above. See also Taube's characteristics of "Western" and "Eastern" (that is Russian) mathematics: "Dualizm Zapada i triedinstvo Vostoka," *MT* 10 (1910), 114–118.

17 See also Michael Hagemeister, "Wiederverzauberung der Welt: Pavel Florenskijs Neues Mittelalter," in N. Franz, M. Hagemeister, and F. Haney, eds, *Pavel Florenskij – Tradition und Moderne. Beiträge zum Internationalen Symposium an der Universität Potsdam, 5. bis 9. April 2000* (Frankfurt am Main, Berlin, Bern, Bruxelles, New York, Oxford, Wien: Peter Lang, 2001), 30; Michael Hagemeister, "Pavel Florenskij und der Ritualmordvorwurf," in Michael Hagemeister and Torsten Metelka, eds, *Appendix 2. Materialien zu Pavel Florenskij* (Berlin u. Zepernick: Kontext, 2001), 71.

18 This use of names is Belyĭ's frequent device. To cite but one example among those which have not been deciphered, "Chukholka" from Belyĭ's earlier novel *The Silver Dove* (1909) is a combination of "Chulkov" (Belyĭ's enemy of that time, the poet and artless theoretician of the so-called "mystical anarchism" G.I. Chulkov) and "Tucholka" (an obscure writer of unsophisticated tracts on occultism, magic etc., S.V. Tukholka).

19 In the course of the 1930s campaign against the Moscow mathematicians Pythagoreans were not forgotten. The Soviet mathematician Mikhail

DOI: 10.1057/9781137338280

Khrisanovich Orlov (1900–1936) devoted most of his Ukrainian booklet on mathematics and religion to a severe critique of the Moscow "school" (I have found this booklet following the quoted article by Eugene Seneta "Mathematics, Religion, and Marxism in the Soviet Union in the 1930s'," *Historia Mathematica* 31: 3 (2004), 349–353). Before addressing the subject of the "school," Orlov writes of the "reactionary mathematics" of earlier times, and in particular, of Pythagoreans characterized as a conservative political group. See M. Orlov, *Matematika i religiïa* (Partvidav "Proletar," 1933), 12.

20 "Nebol'shogo rostochka sukhaîa figurka," "malen'kiǐ," "legkaîa" i "nesolidnaîa" pokhodka, "netopyr'" (Andreǐ Belyǐ, *Nachalo veka* (Moskva: Khudozhestvennaîa literatura), 1990, 157–158). As indicated in the commentaries to *Petersburg*, the similarity with a bat links Ableukhov to Konstantin Pobedonostsev, who was then widely regarded as a symbol of the strong state and reaction (in the contemporary press he was frequently compared with a bat and caricatured as one; S.S. Grechishkin, L.K. Dolgopolov, and A.V. Lavrov, "Primechaniîa," in Andreǐ Belyǐ, *Peterburg* (Sankt-Peterburg: Nauka, 2004), 651, n. 58). What was still more important for Belyǐ, the bat, as well as geometry and speculative thought in general, was traditionally associated with Saturn. It should be emphasized that Belyǐ's portrayal of Petersburg, above which the senator (a close anagram of Saturn) soars like a bat (see Appendix 1), is permeated with Saturnian imagery. Petersburg is represented as the very embodiment of the "reign of Saturn" which had occupied Belyǐ's thought since at least 1909. See Spiritus [Bugaev], "Sem' planetnykh dukhov," *Vesy* 9 (1909), 71; see also n. 69.

21 Belyǐ, *Nachalo veka*, 162. Both Ableukhov and Tikhomirov conduct a diary; both are enemies of the "Jewish press"; and, more importantly, the last of the few encounters with Tikhomirov related by Belyǐ in his memoirs, and the one which apparently made a strong impression on him, occurred in 1911.

22 A.V. Repnikov, "Predislovie," in *Dnevnik L.A. Tikhomirova. 1915–1917* (Moskva: ROSSPĖN, 2008), 10. See also A.V. Repnikov and O.A. Milevskiǐ, *Dve zhizni L'va Tikhomirova* (Moskva: Academia, 2011), 446.
Interestingly, long before the "mystical anarchism" was launched by G.I. Chulkov, Tikhomirov, already a monarchist, warned of its future emergence. Discussing "social mysticism," which informed radical thought, he discerned "in present some outlines of the future *mystical* anarchism," a dangerous outcome of a distorted religious feeling. See Lev Tikhomirov, *Bor'ba veka*, 2nd edn (Moskva, 1896), 11 (italics are by Tikhomirov). Tikhomirov was widely read, so it is probable that he may have actually put the thought and the name of "mystical anarchism" in the mind of Chulkov, whether the latter remembered of this source or not. If that was the case, we have a curious— even if comical, for the actual movement did not live up to Tikhomirov's expectations—example of the latter's alleged subversive activities.

DOI: 10.1057/9781137338280

23 Belyĭ, *Nachalo veka*, 155–161.

24 "*Little*, with a sunken chest <...>, he [Tikhomirov] was dashing around from corner to corner in this nut-brown *cube, wrinkling* his nose and raising his schoulders; <...> without looking at me, he would break the *bat-like* line of his *running* to light a cigarette and—to *rush* into the corner: like a *captive*! [V ètom kube orekhovogo kolorita chesal ot ugla do ugla, vpalogrudyĭ, <...> malen'kiĭ, morshchas' i plechi podniav; <...> ne glîadîa na menîa, preryval svoîu netopyrinuîu liniîu bega: zazhech' papirosu; i—kinut'sîa v ugol: kak plennik!]" (in italics are characteristics which also occur in the descriptions of Ableukhov and his surroundings; Belyĭ, *Nachalo veka*, 157–158).

25 Ippolit Tèn, *Napoleon Bonapart* (Moskva: Musaget, 1912).

26 The publication of this part of Taine's work was immediately related to Medtner's interests, who collected materials on Napoleon (RGB, f. 167, k. 18, ed. khr. 24).

27 H. Taine, *Les origines de la France contemporaine. V. La Révolution. La conquête jacobine*, T. 1 (Paris: Librairie Hachette et Cⁱᵉ, 1904), 23–24.

28 ibid., 27.

29 Let us only note that one of the important sources of Taine was Edmund Burke's *Reflections on the Revolution in France* (1790) in which dangerous abstractedness of revolutionary projects is brought out through numerous geometrical and arithmetical comparisons.

30 In the aftermath of the revolution of 1905, Taine's work was translated both by the Black Hundred (and published as the supplement of the journal *Mirnyĭ trud*, since the middle of 1905 up to the last issue in 1914), and the famous member of the People's Will Herman Lopatin (Ippolit Tèn, *Proiskhozhdenie obshchestvennogo stroîa sovremennoĭ Frantsii*, T. 1 (S.-Peterburg: Izd. M.V. Pirozhkova, 1907). In 1910, while traveling in Sicily, Belyĭ was meditating on the European "geometrical" spirit exemplified in Robespierre and Napoleon. See Andreĭ Belyĭ, *Putevye zametki*, T. 1. *Sitsilîia i Tunis* (Moskva-Berlin: Gelikon, 1922), 101.

31 E.A. Gopius, "Filosofîia 'moskovskoĭ filosofsko-matematicheskoĭ shkoly' i ee otnoshenie k intellektualizmu filosofov XVIII veka i èkonosmicheskomu materializmu K. Marksa," *VFiP* 79: 4 (1905), 554–586.

32 ibid., 585.

33 See Introduction: Belyĭ's Petersburg and Moscow Mathematicians.

34 "The criticism of idealism" in mathematics was claimed to be one of the immediate tasks of Marxism in 1925. See the editor's note to the first article on the subject published by *Pod znamenem marksizma* (Gr. Bammel', "Logistika i dialektika," 3 (1925), 24 n. 1).

35 Andreĭ Belyĭ, *Na rubezhe dvukh stoletiĭ* (Moskva: Khudozhestvennaîa literatura, 1989), 171.

DOI: 10.1057/9781137338280

36 N. Bougaïev, "Les mathématiques et la conception du monde au point de vue de la philosophie scientifique," in *Verhandlungen des ersten internationalen Mathematiker-Kongresses in Zürich vom 9 bis 11 August 1897* (Leipzig: Teubner, 1898), 213.

37 V. Egorshin, *Estestvoznanie, filosofiía i marksizm* (Moskva: Gosizdat RSFSR, Moskovskiĭ rabochiĭ, 1930), 44. The portrayal of the Moscow "school" offered by Egorshin sounded like a police report. He did not forget to make references to the interest in arithmology on the part of Men'shikov (45 n. 2, 47), who had been shot in 1918 and had an established reputation as an extreme reactionary. What must have been especially dangerous for the contemporary mathematicians was the emphasis on Bugaev's wide influence. Egorshin noted that almost every Russian university had Bugaev's pupils teaching in them, and that all Moscow professors had been his pupils (46). Curiously, however, Egorshin seems to have been acquainted with his subject only superficially. Born in 1898, his knowledge of the ideological context of the "school" was limited. Although he described the "school" as the "Black Hundred," he does not appear to have been aware of the extent to which it was really "Black Hundred." His analysis of political implications of arithmology is very inadequate.

38 The following fragment from Élisée Reclus may serve to illustrate this: "Les formules proverbiales sont fort dangereuses, car on prend volontiers l'habitude de les répéter machinalement, comme pour se dispenser de réfléchir. S'est ainsi qu'on rabâche partout le mot de Linné: 'non facit saltus natura.' Sans doute 'la nature ne fait pas de sauts,' mais chacune de ses évolutions s'accomplit par un déplacement de forces vers un point nouveau. Le mouvement général de la vie dans chaque être en particulier et dans chaque série d'êtres ne nous montre nulle part une continuité directe, mais toujours une succession indirect, révolutionnaire, pour ainsi dire." See Élisée Reclus, *L'évolution, la revolution et l'idéal anarchique*, 6-ème éd. (Paris: P.-V. Stock, Éditeur, 1914), 18. This book ran to several Russian editions after the revolution of 1905.

39 To cite but one example close to how Belyĭ described arithmology in his memoirs, "bourgeois" historians of music were attacked by *Pod znamenem marksizma* for their denial of "breaks, leaps, revolutions." See I. Orlov, "Muzyka i klassovaía bor'ba," 10–11 (1925), 198–199. Ivan Efimovich Orlov (1886–1936), a regular contributor to this journal, was a specialist in methodology of natural sciences and mathematics.

40 In his other works of the Soviet period Belyĭ tried to convey the same impression of the revolutionary character of arithmology. Thus, he called it the "sociology of numbers (a doctrine of number community [sotsiologiía chisel (uchenie o chislovykh kollektivakh)]." See Andreĭ Belyĭ, *Ritm kak dialektika i "Mednyĭ vsadnik"* (Moskva: Federatsiía, 1929), 34. Belyĭ was keen

DOI: 10.1057/9781137338280

to draw a parallel between arithmology, which he also called a "doctrine of primacy of the number complex [uchen'e o primate chislovogo kompleksa]" (Belyĭ, *Ritm kak dialektika*, 34), and his treatment of verse. Intending to demonstrate that rhythmical studies of verse were more adequate than metrical ones, the latter were likened to a primitive arithmetical addition, whereas the former to tackling arithmological problems. In studying rhythm one was faced not with simple elements of metric feet, but with a "complex which configures its elements [kompleks, konfiguriruĭushchiĭ èlementy]" (Belyĭ, *Ritm kak dialektika*, 34) On the other hand, rhythm transmitted the "social command," the "sound" sent to poets by society (Belyĭ, *Ritm kak dialektika*, 29–30, 225–232). Thus, rhythm was dictated by a "community," and not by its "elements." Reading this, one gains the impression, which must have been calculated, that both arithmology, as the "doctrine of number community," and Belyĭ's rhythmical studies were a perfect representation, in mathematics and in poetics respectively, of the new spirit of collectivism.

41   This obituary was first referred to in S.M. Polovinkin, "Psikho-aritmo-mekhanik (filosofskie cherty portreta P.A. Nekrasova)," *Voprosy istorii estestvoznaniia i tekhniki* 2 (1994), 112; cf. Oscar B. Sheynin, "Nekrasov's Work on Probability: The Background," *Archive for History of Exact Sciences* 57 (2003), 339.

42   S. Uritskiĭ, "Prof. Pavel Alekseevich Nekrasov: Nekrolog," *Izvestiia* 294 (2329), Decembre 24, 1924, 7.

43   Vladimir Nabokov, *Sobranie sochineniĭ v 4-kh tt.*, T. 4 (Moskva: Pravda, 1990), 293.
There is another curious indication of Nekrasov's success in establishing a good reputation with the new authorities. According to a local newspaper of Sergiev Posad, two streets to be found in this town neighboring Moscow—Nizhne-Nekrasovskaia and Verkhne-Nekrasovskaia—owe their names not to the famous Russian poet, but to our mathematician. In August 1924, when Nekrasov was still alive, Komsomol members of Sergiev, which, before the revolution of 1917, had been an important religious center with Troitse-Sergieva Lavra and Moscow Theological Academy situated there, initiated a renaming of streets of their town. Newly chosen names had to be compatible with the new revolutionary culture. See Aleks Rdultovskiĭ, "Zdravstvuĭte, Pavel Alekseevich!," *Vpered. Munitsipal'naia obshchestvenno-politicheskaia gazeta Sergievo-Posadskogo raiona. Kraevedcheskiĭ vestnik*, July 23, 2011, http://www.vperedsp.ru/statyi/kraeved/?ID=3745.

44   Cf. Belyĭ, *Putevye zametki*, 26.
Although once included in N.K. Krupskaia's Index in the early 1920s, Plato firmly established himself as a "Soviet classic" by 1960–1970ss, when his works were printed in great numbers of copies. See G.Ch. Guseĭnov, "Politicheskiĭ platonism, obnazhaĭushchiĭ i skryvaĭushchiĭ," in *Antichnost' i*

DOI: 10.1057/9781137338280

kul'tura Serebrîanogo veka: K 85-letîïu A.A. Takho-Godi (Moskva: Nauka, 2010), 533–539.
After the revolution Nekrasov is reported to have said that "he made a mistake in his previous 'works': he selected the wrong sign of a square root. When replacing it with the contrary sign, he will be able to prove the need for social revolution" (quoted in Sheynin, "Nekrasov's Work on Probability," 340).

45 Andreĭ Belyĭ, *Mezhdu dvukh revolîutsiĭ* (Moskva: Khudozhestvennaîa literatura, 1990), 283; Mikhail Bezrodnyĭ, "O 'îudoboîazni' Andreîa Belogo," *Novoe literaturnoe obozrenie* 28 (1997), 113, 124 n. 98; cf. Szilard, *Germetizm i germenevtika*, 261–263.

46 See Appendix 1.

47 D. Rusinov, "Khristianskiĭ mir i voinstvuîushchee iudeĭstvo," *MT* 1 (1906), 86.

48 Aleksandr Selîaninov, *Taĭnaîa sila masonstva* (S.-Peterburg, 1911), 284–285. An interesting document records the reaction of an important functionary and probably a Mason, Èduard Nikolaevich Berendts (1860–1930) to this propaganda. In 1911 he published a brochure entitled *Masonstvo ili velikoe tsarstvennoe iskusstvo vol'nykh kamenshchikov kak kul'turoispovedanie* [Masonry or the great royal art of the free Masons as a cultural creed] (S.-Peterburg, 1911). (The copy that I have consulted in the Russian Public library was previously held in the library of Duma). The Masonry is represented here as a profoundly Aryan phenomenon (17); the core of its program is claimed to be purely cultural, its political object being limited to the struggle against the Jews and Jesuits.

49 Echoes of this view could even be found in the history of mathematics. Thus, Ferdinand Hoefer called Pythagoreans "ces antiques francs-maçons" (*Histoire des mathématiques: depuis leurs origines jusqu'au commencement du dix-neuvième siècle* (Paris: Librairie Hachette et Cⁱᵉ, 1874), 92). Cf. in Bulwer Lytton (to be quoted by Lewes): "the political designs of his [Pythagoras'] gorgeous and august philosophy, only for a while successful, left behind them but the mummeries of an impotent freemasonry, and the enthusiastic ceremonies of half-witted ascetics" (Edward Bulwer Lytton, *Athens: Its Rise and Fall,* Vol. II (Leipzig: Bernard Tauchnitz, 1843), 240; Lewes, *Istorïïa filosofii ot nachala ee v Gretsii do nastoîashchikh vremen* (S.-Peterburg, 1865), 22). If, as suggested above, the political activity of Pythagoreans was better remembered at that time, this must have contributed to their "Masonic" reputation, and vice versa.

50 See, for example: F.-T. B.-Clavel, *Histoire pittoresque de la franc-maçonnerie et des sociétés secrètes anciennes et modernes* (Paris: Pagnerre, Éditeur, 1843), 172. A certain Karl Oppl thus admonished his fellow Masons: "Ne soyons-nous jamais inférieurs aux nobles Pythagoriciens!" See Karl Oppl, *Pythagore et la*

DOI: 10.1057/9781137338280

Fran-Maçonnerie (Francfort s.M.: Ferdinand Bosell, 1861; reprinted: Nîmes: Lacour, 2000), 47.

51   Nekrasov, "Moskovskaîa filosofsko-matematicheskaîa shkola," 100. Cf.: "Such an architector is good, and such masons and carpenters [are good] who build every part of a building according to their calculations [po raschetu]. This rule is still more applicable to the social architectonics [arkhitektonike] which normalizes [normiruet] mutual relations in the field of political and social life" (id., "Logika mudrykh lîudeï i moral," 919).

52   Plato's *Timaeus* may have been Nekrasov's immediate source. Since Timaeus was a Pythagorean, it is possible that for Nekrasov "world soul" was not so much the current term of contemporary writings (see Chapter 3, n. 20), as that of the venerable tradition of which he felt himself a part and which made him responsive to current ideas on the subject.

53   Nekrasov, "Moskovskaîa filosofsko-matematicheskaîa shkola," 105.

54   ibid., 192.

55   Taube, *Svod osnovnykh zakonov myshleniîa*, 137–140.

56   See M. Vashutin [M.F. Taube], *K Vozrozhdeniiu Slavîano-Russkogo Samosoznaniîa* (Petrograd, 1912), 113–118.

57   RGALI, f 53, op. 1, ed. khr. 100, L. 123; Belyï and Ivanov-Razumnik, *Perepiska*, 341. Belyï was working on *Istoriîa* from the middle of the 1920s till the early 1930s. For the details concerning his work on this treatise, see Monika Spivak, "Andreï Belyï v rabote nad traktatom *Istoriîa stanovleniîa samosoznaîushcheï dushi*," *Russian Literature* 70: 1/2 (2011), 1–19.

58   RGALI, f. 53, op.1, ed. khr. 80, L. 7 – 70b.

59   ibid.

60   See Abel-Rémusat, *Mémoire sur la vie et les opinions de Lao-Tseu, philosophe chinois du VI.e siècle avant notre ère, qui a professé les opinions communément attribuées à Pythagore, à Platon et à leurs disciples* (Paris, 1823).

61   Aug. Gladisch, *Einleitung in das Verständniss der Weltgeschichte. Erste Abtheilung: Die alten Schinesen und die Pythagoreer* (Posen, 1841), 204–205.

62   RGB, f. 167, k. 18, ed. khr. 9, L. 31; the same is repeated in RGB, f. 167, k. 18, ed. khr. 12, L.17. Cf. Houston Stewart Chamberlain, *The Foundations of the Nineteenth Century*, trans. John Lees, 2nd ed., Vol. I (London – New York: The Bodley Head; John Lane Company, 1912), 47, 54–57.

63   This observation was common knowledge at that time. One encounters it, either accepted or refuted in works concerning various subjects. See, for example, Camille Flammarion, *L'inconnu et les problèmes psychiques* (Paris: Ernest Flammarion, Éditeur, 1900), 2–3; ÎA. Veïnberg, *Nikolaï Kopernik i ego uchenie* (S.-Peterburg, 1873), 10–16.

64   Ilona Svetlikova, "Kant-semit i Kant-ariets u Belogo," *Novoe literaturnoe obozrenie* 93: 5 (2008), 74.

65   ibid., 87.

DOI: 10.1057/9781137338280

66 A similar train of thought is to be found in *Putevye zametki*, in which the "geometrical" spirit of Europe is identified with that of the chaotic Eastern ornament (101).

67 Belyĭ refers to this story in his memoirs (Belyĭ, *Nachalo veka*, 70). A detailed analysis of this theme is to be found in I. ÎU. Svetlikova, "'Edinorog, probodaĭushchiĭ rytsarîa': iz kommentariev k 'Peterburgu' Andreîa Belogo," in A.F. Nekrylova, ed., *Zelenyĭ Zal-3: al'manakh / Rossiĭskiĭ institut istorii iskusstv* (Sankt-Peterburg: RIII, 2013), 73–79.

68 Andrei Bely, *Petersburg,* trans. Robert A. Maguire and John E. Malmstad (Bloomington – London: Indiana University Press, 1978), 20.

69 Maria Carlson pointed out that, following the European iconological tradition, the unicorn on the Ableukhov coat of arms symbolized Christ; it promised the spiritual rebirth occuring at the end of the novel. See "The Ableukhov Coat of Arms," in Boris Christa, ed., *Andrey Bely Centenary Papers* (Amsterdam: Verlag Adolf M. Hackert, 1980), 157–170. In the above mentioned *Sem' planetnykh dukhov* the "reign of Saturn" is to be ruined by the "Lamb" (see n. 20).

Both killing and reviving, the unicorn placed on the Ableukhov coat of arms (and, in a sense, on that of Belyĭ himself) is an appropriate emblem of the dialectics permeating the novel and Belyĭ's thought.

70 In 1912, while writing the novel, Belyĭ became Rudolf Steiner's disciple. The following year the latter abandoned Theosophical society, but not the ideas we discuss.

71 Conan-Doyle, *Strannoe prividenie*, in his *Polnoe sobranie sochineniĭ*, Kn. 3 ([Sankt-Peterburg]: zhurnal "Priroda i Lîudi," 1909), 241–256.

72 Bely, *Petersburg*, 65.

73 The image of Mongol hordes frightening anti-Semitic Dudkin was inspired by Rightist propaganda (see Chapter 3, Sections "1911 in the history of the extreme Right," and "1911 in *Petersburg*"; it will be recalled that Lippanchenko looks like a Mongol). In some contemporary sources familiar to Belyĭ the worship of the sun was perceived as essentially Aryan. Here is what Schuré wrote in the chapter on Pythagoras: "L'adoration de l'homme aryen se porta dès l'origine de la civilisation vers le soleil comme vers la source de la lumière, de la chaleur et de la vie." See Édouard Schuré, *Les Grands Initiés: Esquisse de l'histoire secrète des religions* (Paris: Perrin et Cⁱᵉ, Libraires-Éditeurs, 1889), 293.

74 In the first version Dudkin's words were slightly different: "And on that day the final Sun will rise in radiance over my native land: this will be our Lord, Christ [to Gospod' nash, Khristos]" (L.K. Dolgopolov, "Tekstologicheskie printsipy izdaniîa," in Belyĭ, *Peterburg*, 630). Having removed this precision on the meaning of the sun, which, as Dolgopolov correctly pointed out, rendered the whole passage more polysemantic, Belyĭ's aim was rather to

DOI: 10.1057/9781137338280

make a clearer hint at the Pythagorean constituent of Dudkin's character, than to create a vague allegorical message.

As a visual counterpart of Dudkin's prayer, one may refer to the painting by Fedor Bronnikov *The Hymn of Pythagoreans in Praise of the Rising Sun* (1869), in which Pythagoreans look like early Christians, which reflects an old tradition of considering the former as the precursors of the latter. For the scholarly expression of this tradition, see Isidore Lévy, *La Légende de Pythagore de Grèce en Palestine* (Paris: H. Champion, 1927).

75   See, in particular, Andreï Belyĭ, "Krugovoe dvizhenie. (Sorok dve arabeski)," *Trudy i dni* 4–5 (1912), 51–73; id., *Putevye zametki*, 100–102. The fact that Belyĭ attached geometrical emblems to different outlooks and manners of thought confirms that the ideological use of geometry in the novel was not a chance and semi-conscious artistic device; it was fully deliberate.

DOI: 10.1057/9781137338280

Conclusion

Abstract: *The ideology of the ultra-Right proves to be the immediate framework for the obscure works of the Moscow "school." The attempt to make mathematics part of the Black Hundred discourse sheds new light on the intellectual atmosphere on the eve of the revolutions in Russia.*

Svetlikova, Ilona. *The Moscow Pythagoreans: Mathematics, Mysticism, and Anti-Semitism in Russian Symbolism.* New York: Palgrave Macmillan, 2013. DOI: 10.1057/9781137338280.

One of the aims of the foregoing pages was to describe a reactionary and ultimately obscurantist ideology that claimed to be founded on mathematical grounds. It was given utterance in a number of works some of which were of such complexity and obscurity that they were later suspected of being the product of an unhinged mind.

This ideology originated in the atmosphere of the nineteenth-century pursuit of an all-embracing system of knowledge. N. V. Bugaev, the instigator and founder of some of the key features of the Moscow philosophic-mathematical school, was, in various periods of his life, an assiduous reader of positivistic writers who purported to offer such a system, of contemporary psychology which was viewed as a new fundamental discipline of universal import; of philosophers, among whom Leibniz' versatility held a particular fascination for him; and of theosophists, who treated of everything. Interested in the history of mathematics, he was also acquainted with the view of mathematics that accorded it predominance among the sciences and humanities. He came up with the conception of arithmology as a foundation of a new system of thought, in which mathematics would expand to embrace traditionally non-mathematical subjects.

Arithmology, related to the continuity/discontinuity problem, the discussions of which were often charged with religious and political implications, became impregnated with ideological significance. Bugaev's followers believed that the new "arithmological" framework could corroborate Christian faith using mathematics. Suggesting that determinism was not absolute, and miracles, as related in the Christian tradition, were actually possible, arithmology acquired a prominent religious aspect.

Owing to Bugaev's disciple, P. A. Nekrasov, who established a parallel between right-wing demands that the Russian Tsar be free from constitutional laws and the "arithmological" assumption that the power of natural laws was not absolute, mathematics came to be promoted as a scientific basis for ultra-conservative doctrines.

Bugaev's conception of arithmology influenced Nekrasov's views of theory of probability, which he regarded as a Christian groundwork for all departments of human activity. As right-wing propaganda intensified after the failure of the Russian revolution of 1905, attributing broad applicability and Christian spiritual significance to theory of probability made it compatible with the Slavophil philosophy of "whole knowledge" and the Russian national spirit in general. Due to its use in economics, the Christian and national connotations of theory of probability

DOI: 10.1057/9781137338280

received an intensive anti-Semitic coloring. Theory of probability started to be perceived as a weapon against the impending danger disclosed by the *Protocols of the Elders of Zion*, which haunted the imagination of the Moscow "school."

Writings related to the Moscow philosophic-mathematical school are both unique, merging professional mathematics with religion and politics in a way highly unusual for that time, and very suggestive of the contemporary political climate, conveying through its very uniqueness the pervasive character of anti-Semitism more clearly than other better known facts do.

Having been brought up among the books of his father's "universal library" and strongly marked by his influence, Belyĭ exhibited the same avidity for various branches of knowledge as Bugaev and some of his disciples did. The nature of Belyĭ's religious quest that brought him within the sphere of theosophists and later Rudolph Steiner was not dissimilar to Bugaev's interest in theosophy. Likewise, Belyĭ's "mystical anti-Semitism," which is crucial for his novel *Petersburg* and his collection of essays *Symbolism*, was not unrelated to Bugaev's political views, equally characteristic of his "school."

Thus, departing from the motif of geometry in *Petersburg*, this book has endeavored to offer new approaches to understanding the message of the Moscow philosophic-mathematical school, and to the meanings of one of the past century's literary masterpieces. Geometry in Belyĭ's novel guides us to important aspects of the history of both Russian mathematics and Russian symbolism, shedding new light on the frightful confusion that overpowered minds on the eve of the revolutions in Russia. Observing professional mathematicians in search of fanciful weapons against imaginary dangers gives one a valuable snapshot of the intellectual atmosphere hastening the fall of the old regime.

*   *   *

At one point of my research, Omry Ronen, who was then in Saint-Petersburg, after having read the manuscript of a preliminary version of this study, wrote at the back of a page:

> Look <...> the famous Norwegian mathematician and the head of various research centers on the problems of peace and war, Johan Galtung, has declared that the mass killing in Norway was organized by the Mossad and the Jews, and adviced to read the *Protocols of the Elders of Zion*.[1] <...> the

DOI: 10.1057/9781137338280

hatred to the Jews has been the basis of the mostly diverse ideologies of the last century and a half or more [v osnove samykh raznoobraznykh ideologiĭ poslednikh polutora ili bolee vekov lezhit nenavist' k evreĭam]. This is the very hydra, sometimes it hides itself, sometimes it sticks its heads out [Èto nastoĭashchaĭa gidra i est', ona to prĭachetsĭa, to tut i tam vysovyvaet golovy]. Hence the topicality of this subject and the fear <...> before it [Otsĭuda aktual'nost' Vasheĭ temy i strakh <...> pered neĭ].

Note

1   See Ofer Oderet, "Pioneer of Global Peace Studies Hints at Link between Norway Massacre and Mossad," Haaretz, April 30, 2012 in: http://www. haaretz.com/news/diplomacy-defense/pioneer-of-global-peace-studies-hints-at-link-between-norway-massacre-and-mossad-1.427385; cf. https://www.transcend.org/galtung/statement-may-2012/.

DOI: 10.1057/9781137338280

Appendix 1: Geometrical Meditations of the Senator

The carriage was flying toward Nevsky Prospect.

Apollon Apollonovich Ableukhov was gently rocking on the satin seat cushions. He was cut off from the scum of the streets by four perpendicular walls. Thus he was isolated from people and from the red covers of the damp trashy rags on sale right there at this intersection.

Proportionality and symmetry soothed the senator's nerves, which had been irritated both by the irregularity of his domestic life and by the futile rotation of our wheel of state.

His tastes were distinguished by their harmonious simplicity.

Most of all he loved the rectilineal prospect; this prospect reminded him of the flow of time between the two points of life.

There the houses merged cubelike into a regular, five-story row. This row differed from the line of life only in one respect: this row had neither beginning, nor end <...>.

Inspiration took possession of the senator's soul whenever the lacquered cube cut along the line of the Nevsky: there the numeration of the houses was visible. And the circulation went on.

    <...>

While gazing dreamily into that illimitability of mists, the states-man suddenly expanded out of the black cube of the carriage in all directions and soared above it. And he wanted the carriage to fly forward, the prospects to fly to meet him—prospect after prospect, so that the entire spherical surface of the planet should be embraced, as in serpent coils, by blackish gray cubes of houses; so that all the earth, crushed by prospects, in its lineal cosmic flight should intersect, with its rectilineal principle [zakonom], unembraceable infinity; so that the network of parallel prospects,

DOI: 10.1057/9781137338280

intersected by a network of prospects, should expand into the abysses of the universe in planes of squares and cubes: one square per "solid citizen," so that…[1]

After the line, the figure which soothed him more than all other symmetries was the square.

At times, for hours on end, he would lapse into an unthinking contemplation of pyramids, triangles, parallelepipeds, cubes, and trapezoids.

While dwelling in the center of the black, perfect, satin-lined cube, Apollon Apollonovich reveled at length in the quadrangular walls. Apollon Apollonovich was born for solitary confinement. Only his love for the plane geometry of the state had invested him in the polyhedrality of a responsible position.[2]

Notes

1   Cf. the wording of P.A. Nekrasov's reflections on the "magnitude of the territory of the great state, which is to be covered in all points by the pacifying action of good laws [obshirnost' territorii velikogo gosudarstva, kotoruîu nuzhno vo vsekh punktakh pokryt' umirotvorîaîushchim deîstviem blagikh zakonov]" (*Gosudarstvo i Akademiîa* (Moskva, 1905), 73); cf. also a similar passage in the *Moscow Philosophic-Mathematical School*, in which Nekrasov writes on the "social architectonics [sotsial'naîa arkhitektonika]" which must "cover [pokryt']" the society by the "network [setîu]" of the institutions carrying laws into effect (P.A. Nekrasov, "Moskovskaîa filosofsko-matematicheskaîa shkola i ee osnovateli," *MS* 25: 1 (1904), 191).

2   This fragment is taken from one of the first sections of the novel entitled "Squares, Parallelepipeds, Cubes" describing Ableukhov's journey to the mysterious "Office" of which he is a head (Andrei Bely, *Petersburg*, trans. Robert A. Maguire and John E. Malmstad (Bloomington – London: Indiana University Press, 1978), 10–11).

DOI: 10.1057/9781137338280

Appendix 2: Rhythm

Commenting on *Symbolism*, it is important to note that the articles on rhythm, which, according to Belyĭ's own words, formed its core,[1] were not isolated from the context of "Musaget's" anti-Semitism. There was no immediate connection between Belyĭ's main discovery—the definition of the difference between the poetic rhythm and meter[2]—upon which these articles were based, and the fit of Judaeophobia that struck him at that time. Nevertheless, putting his studies on rhythm in their historical context, the ideas of Aryan renaissance preached by "Musaget" turn out to be relevant.

Medtner touched upon the questions of rhythm in his articles on music.[3] His views on the subject were, in particular, expressed in his collected articles *Modernizm i muzyka* ([Modernism and Music], 1912) which came out in "Musaget" (the articles dated to 1907–1910, and the appendix to 1911). This volume was directed against the Jewish musicians and composers who, as the author believed, posed a mortal threat to the Aryan music and had, therefore, to be exposed as such.[4] He disparaged their works in various ways, rhythm being one of the chief targets. Jewish composers and conductors (or those whom he, for some reason, regarded as such) were characterized as profoundly "arythmical"; they suffered from "rhythmical infirmity [ritmicheskaĭa nemoshch']"[5] and an inability to create new rhythms[6]; they had mere "metrical skills" which could not replace the "natural musical rhythmicity [prirodnuĭu muzykal'nuĭu ritmichnost']";[7] Wagnerian rhythms were beyond their powers.[8]

The culmination of the "rhythmical" part of Medtner's obsession comes in the appendix, which shows the author fully sharing in the contemporary vogue for the concept (or rather the word) of rhythm. The latter was charged with intellectual and emotional implications which elude us now, but must have been powerful then, for there was a veritable "delirium rhythmicum." Although this particular expression was coined to designate a widespread passion for rhythmical gymnastics,[9] it appears to have been of much wider relevance.

> Rhythm is the magnificent course of the lifegiving stream of nature [zhivotvorîashcheĭ strui prirody] in the universe. Rhythm is the law of the movement of the celestial spheres, and the ancient thinkers of Ellas were feasting on its harmony in a sacred ecstasy. Rhythm governs the solar world; it generated time, and it moved planets into space... The army marches under its command; the ballerina gets her charms from it, it dominates music, and it commands the all-mighty human speech.[10]

This quotation is taken from a book on Russian verse, written after *Symbolism*, but it perfectly exemplifies the tendency originating much earlier. As was noted by the reviewer of the book, the word "rhythm" did not mean anything except in cases when terminologically defined.[11]

In comparison with Belyĭ's lucidity in his treatment of rhythm in *Symbolism*, Medtner's observations on the subject do not essentially differ from those just quoted. "Rhythm is the primary joy [pervichnaîa radost']. Rhythm is the soul of music... <...> all the elements of music can be reduced to rhythm"[12]; "Rhythm is the primary creative element [tvorcheski-pervichnyĭ èlement], <it is> something individual, inborn, free in its movement, irrational <...>."[13]

Characteristically, close to the very beginning of the appendix on rhythm, there is a reference to an Indian source.[14] The discussion of rhythm was incorporated into Medtner's racial discourse, making rhythm a natural property of free and creative Aryan spirit, an expression of racial and national individuality, incompatible with the "international" character of the Jewish music.

The fact that Belyĭ attained a remarkable degree of precision in his studies does not mean that he was unaffected by the current tendency to regard rhythm as pertaining to the very essence of things or to the very essence of the various problems with which he was dealing. Especially in his later works, rhythm is explicitly and heavily charged with philosophical and ideological meanings. Thus, his *Ritm kak dialektika i "Mednyĭ vsadnik"*, among other things, is representative of his

DOI: 10.1057/9781137338280

attempts to find his place within the Soviet framework, in this case, as the title indicates, by affiliating his rhythmical studies to the Hegelian philosophy.

At that point Belyĭ was trying to represent both *Petersburg* and his works on rhythm as profoundly in tune with the new revolutionary times.[15] In neither case was this difficult, not only because both were revolutionary in a sense of being very original. In both the opposition was present between a force of cultural inertia (poetical meter; political conservatism of the senator) and a creative impulse (poetical rhythm, which breaks the monotony of convenient meters; the break of the "circular motion" achieved by Ableukhov the younger in the last pages of the novel). Sclerosis is both a characteristic of meter in Belyĭ's studies of the Soviet period[16] and a hereditary feature of Ableukhovs, a sort of recurrent physical "rhyme" to their Oriental immobility.

The point important for a historian is, however, that at the time of writing *Petersburg*, the conflict between the principle of cultural and social creation, on the one hand, and that of inertia and stagnation, on the other, was perceived by Belyĭ not in social, but in racial terms. He did not think of class struggle: it was the fight between the Aryan and Semitic principles which, in the framework of ideology adopted by "Musaget," defined political and cultural history.

This is obvious from the analysis of *Petersburg*. This is no less obvious from the analysis of the "symbolic worldview" in *Symbolism* (see above). Looking at the articles on rhythm, one would have seen a theoretical pronouncement having a direct bearing on the problem of Aryan culture, of which rhythm was an essential feature. That Belyĭ was not unaffected by such views is clear from his notorious article *Shtempelevannaîa kul'tura* ([Stamped culture], 1909), in which he formulated the necessity of equal rights for the Jews in the following terms:

> <...> this would draw their attention away from literature and arts towards the domain of political life: the Jew as a statesman would be useful, and necessary; the Jews are statesmen by their nature [evrei po prirode gosudarstvenniki]; whereas all the true breath of the Aryan culture is beyond state, being free and rhythmical [vsîakoe zhe istinnoe dykhanie ariĭskoĭ kul'tury vnegosudarstvenno, svobodno, ritmichno].[17]

This does not elucidate Belyĭ's theoretical contribution to the study of verse in *Symbolism*. Yet *Symbolism* itself becomes more coherent: the articles on rhythm associated with "the true breath of Aryan culture" formed the natural core of a volume laying the foundations for an Aryan world-view.

DOI: 10.1057/9781137338280

Notes

1   In *Retrospective Diary* Belyĭ writes that in March 1909 after intensive work on poetic rhythm he got "a solid core [tverdyĭ kostĭak] of material for *Symbolism*" (*Rakkurs k dnevniku*, RGALI, f. 53, op. 1, ed. khr. 100, L. 47).

2   "*By the rhythm of a poem we understand a symmetry in its deviations from meter, that is a certain complex uniformity of deviations*" (Andreĭ Belyĭ, *Simvolizm* (Moskva: Musaget, 1910), 396; italics are by Belyĭ). On Belyĭ's studies of rhythm, see M. Spivak, M. Odesskiĭ, "'Èto popytka skazat' o zheste ritma vne stikhovedcheskoĭ laboratorii...'. Ob odnoĭ neopublikovannoĭ state Andreĭa Belogo," *Voprosy literatury* 2 (2010), 224–245, and the references there given.

3   Belyĭ and Medtner discussed rhythm as early as 1908, when the latter noted in the entry of his diary dated August 10: "Das einzige Interessante was diesmal Bugaev besprochen hat war seine Theorie des Rhythmus, die in manchen mit meiner übereinstimmt" (RGB, f. 167, k. 22, ed. khr. 12, L. 38).

4   Medtner's brother was the composer and musician Nikolaĭ Karlovich Medtner (1880–1951), whose music he propagated, desiring to protect it from the "unfair" competition.

5   Vol'fing [È.Medtner], *Modernizm i muzyka* (Moskva: Musaget, 1912), 9; cf. ibid., 5.

6   ibid., 57.

7   ibid., 320, n. 2.

8   ibid., 23.

9   Sergeĭ Volkonskiĭ, *Chelovek na stsene* (S.-Peterburg: Izd. "Apollona," 1912), 162 n.

10   D.G. Gintzburg, *O russkom stikhoslozhenii. Opyt issledovaniĭa ritmicheskogo stroĭa stikhotvoreniĭ Lermontova* (Petrograd, 1915), 34–35. The "rhythmical" vogue of the beginning of the past century was briefly outlined in Ilona Svetlikova, "Bergson v 'Novom LEFe'," in Willem G. Weststeijn, ed., *Delo avangarda. The case of the Avant-Garde* [materials of the conference held in Amsterdam, May 7–10, 2006] (Amsterdam: Uitgeverij Pegasus, 2008), 517–519. See also Ĭuriĭ Orlitskiĭ, "Ritm v filosofsko-èstèticheskikh iskaniĭakh Andreĭa Belogo i v *Istorii stanovleniĭa samosoznaĭushcheĭ dushi* v kontekste iskaniĭ ego vremeni," *Russian Literature* 70: 1/2 (2011), 175–194.

11   S. Bobrov, *Zapiski stikhotvortsa* (Letchworth – Herts – England: Prideaux Press, 1973), 71.

12   Vol'fing, *Modernizm*, 355.

13   ibid., 357.

14   ibid., 355.

15   See Chapter 5, n. 40.

16   Meter is identified with "historical canon" and "class sclerosis [klassovoĭ sklerotizatsii]" (Belyĭ, *Ritm kak dialektika*, 25; cf. ibid., 19, 220).

DOI: 10.1057/9781137338280

17   Boris Bugaev, "Shtempelevannaĭa kul'tura," *Vesy* 9 (1909), 77–78. In Belyĭ's
     second letter to Alexander Blok, dating January 6, 1903, there is the following
     observation: "Rhythm—as a repetition of the time pulse—is connected with
     the idea of the Eternal Return <…> [Ritm—kak povtornost' vremennogo
     pul'sa—svĭazan s ideeĭ Vechnogo Vozvrashcheniĭa]" (Andreĭ Belyĭ and
     Aleksandr Blok, *Perepiska, 1903 – 1919* (Moskva: Progress-Pleĭada, 2001), 25).
     The idea of the eternal return was tightly associated with Indian philosophy.

DOI: 10.1057/9781137338280

Appendix 3: "The Sentiment of Nature"

<...> having found himself on one rare occasion in the flowering bosom of nature, Apollon Apollonovich saw: the flowering bosom of nature. For us this bosom would immediately break down into its characteristics: into violets, buttercups, pinks; the senator would again reduce the particulars to a unity. We would say, of course:

"There's a buttercup!"

"There's a nice little forget-me-not..."

But Apollon Apollonovich would say simply and succinctly:

"A flower..."

Just between us: for some reason, Apollon Apollonovich considered all flowers the same, bluebells.[1]

This curious trait, another expression of the senator's bent for abstract thinking, is not entirely of Belyĭ's invention. The senator's blindness to the "flowering bosom of nature" was prompted by the discussion of the so-called sentiment of nature. Another echo of this discussion dating from 1911 is to be found in Vladimir Zhabotinskiĭ, who makes "a Russian" of his dialog *An Exchange of Compliments* say: "<...> the colors of nature, of the sky, of the sea, of the leaves—all this is ignored, as if it does not exist, [as if] it is not necessary, not interesting for the dry, calculating monotonous Jewish spirit."[2]

Such notions were supported by scholarly literature, relating to a trend which seems to have originated with Alexander von Humboldt (1769–1859). The second volume of his influential and popular *Kosmos* (1847) was devoted to

DOI: 10.1057/9781137338280

the analysis of Naturgefühl. Humboldt stated therein that the character of the sentiment of nature varies with times and races, and, in particular, compared that of the Hindus with that of the Jews. Observing the differences between the Aryan and Semitic descriptions of nature, he was, however, equally sympathetic with them.[3] This was not the case of all the writers who followed him. Apollon Apollonovich, bent on reducing "the particulars to a unity," displayed an allegedly characteristic feature of the Semitic mind, as notably formulated in one of the most prolific authors on the subject. In his early book *Die Entwickelung des Naturgefühls im Mittelalter und in der Neuzeit* (1888), with which Florenskiĭ opens the bibliographical list of works concerning the sentiment of nature in his *Pillar and Ground of the Truth*,[4] the German literary critic Alfred Biese (1856–1930) draws a sharp contrast between the Aryan attitude with that of the Jews.[5] The first are great lovers of nature; they love it for its own sake, they are full of sympathy towards it, hence their representations of it are rich and detailed. On the other hand, for the Jews, nature, being a manifestation of God, is nothing but "dust" in itself; it is "leblos," and they know no Aryan sympathy towards it, nor do they perceive phenomena in their concreteness and specificity: "der Blick ins Unbegrenzte hemmt das liebevolle Ergründen des Einzelnen, die Erfassung der individuellen Gestalt und der Wesenheit der Erscheinungen."[6]

Notes

1   Andrei Bely, *Petersburg*, trans. Robert A. Maguire and John E. Malmstad (Bloomington – London: Indiana University Press, 1978), 21.

2   Vl. Zhabotinskiĭ, *Fel'etony* (S.-Peterburg, 1913), 184.

3   "Tiefes Naturgefühl spricht sich in den ältesten Dichtungen der Hebräer und Inder aus: also bei Volkstämmen sehr verschiedener, semitischer und indogermanischer Abkunft" (Alexander von Humboldt, *Kosmos. Entwurf einer physischen Weltbeschreibung*, Bd. 2 (Stuttgart und Tübingen: J.G. Cotta'scher Verlag, 1847), 7; cf. Humboldt, *Kosmos*, 39–50).

4   P.A. Florenskiĭ, *Stolp i utverzhdenie istiny* (Moskva: Pravda, 1990), T. 1 (II), 736, n. 487. This note is related to his statement that the "sentiment of nature" appeared in the world with Christianity (*Stolp i utverzhdenie istiny*, T. 1 (I), 275).

5   Alfred Biese, *Die Entwickelung des Naturgefühls im Mittelalter und in der Neuzeit* (Leipzig: Verlag von Veit & Comp., 1888), 8–16.

DOI: 10.1057/9781137338280

6   ibid., 16. The attention of the senator in the course of his "geometrical" voyage in the beginning of the novel is perhaps not by accident attracted by "illimitability"—of a sort which goes well with Belyi's kind of mockery: we see Apollon Apollonovich "gazing dreamily into <the> illimitability of mists" (Bely, *Petersburg*, 11).

DOI: 10.1057/9781137338280

Select Bibliography

Alekseev, V.G. *Gerbart, Strumpel i ikh pedagogicheskie sistemy*. ÎUr'ev, 1907.

—— *K voprosu ob obrazovatel'nom znacheniii kursa teorii veroîatnosteĭ dlîa sredneuchebnykh zavedeniĭ*. ÎUr'ev, 1914.

—— *Plody vospitatel'nogo obucheniîa v dukhe Komenskogo, Pestalotstsi i Gerbarta*. ÎUr'ev, 1906.

Alexejeff, W.G. "Über die Entwickelung des Begriffes der höheren arithmologischen Gezetsmässigkeit in Natur- und Geisteswissenschaften." *Vierteljahrsschrift für wissenschaftliche Philosophie u. Soziologie* 3: 1 (1904), 73–92.

Andreev, A.V. "Teoreticheskie osnovy doveriîa (shtrikhi k portretu P.A. Nekrasova)." *IMI. Vtoraîa seriîa* 39: 4 (1999), 98–113.

Busev, V.M. "Shkol'naîa matematika v systeme obshchego obrazovaniîa 1918–1931 gg." *IMI. Vtoraîa seriîa* 47: 12 (2007), 68–97.

Bely, Andrei. *Petersburg*. Translated by Robert A. Maguire and John E. Malmstad. Bloomington – London: Indiana University Press, 1978.

Belyĭ, Andreĭ. "Krugovoe dvizhenie. (Sorok dve arabeski)." *Trudy i dni* 4–5 (1912), 51–73.

—— *Na rubezhe dvukh stoletiĭ*. Moskva: Khudozhestvennaîa literatura, 1989.

—— *Nachalo veka*. Moskva: Khudozhestvennaîa literatura, 1990.

—— *Peterburg*. Sankt-Peterburg: Nauka, 2004.

—— *Putevye zametki. T. 1. Sitsiliîa i Tunis*. Moskva-Berlin: Gelikon, 1922.

_____ *Ritm kak dialektika i "Mednyĭ vsadnik."* Moskva: Federatsiia, 1929

_____ *Sobranie sochineniĭ.* Moskva: Respublika, 1997.

_____ *Simvolizm.* Moskva: Musaget, 1910.

Belyĭ, Andreĭ, and Blok, Aleksandr. *Perepiska, 1903–1919.* Moskva: Progress-Pleîada, 2001.

Belyĭ, Andreĭ, and Ivanov-Razumnik. *Perepiska.* Sankt-Peterburg: Atheneum; Feniks, 1998.

Berglund, Krista. *The Vexing Case of Igor Shafarevich, a Russian Political Thinker.* Basel: Springer, 2012.

Bezrodnyĭ, Mikhail. "Iz istorii russkogo germanofil'stva: izdatel'stvo 'Musaget.'" *Issledovaniia po istorii russkoĭ mysli. Ezhegodnik za 1999 god* (Moskva: OGI, 1999), 157–198.

_____ "O 'iudoboîazni' Andreîa Belogo." *Novoe literaturnoe obozrenie* 28 (1997), 100–125.

Biese, Alfred. *Die Entwickelung des Naturgefühls im Mittelalter und in der Neuzeit.* Leipzig: Verlag von Veit & Comp., 1888.

Bobynin, V. "Matematika." In *Èntsiklopedicheskiĭ slovar'.* Edited by F.A. Brokhaus, and I.A. Efron. T. XVIII^A , 781–795. S.-Peterburg, 1896.

Bougaïev, N. "Les mathématiques et la conception du monde au point de vue de la philosophie scientifique." In *Verhandlungen des ersten internationalen Mathematiker-Kongresses in Zürich vom 9 bis 11 August 1897*, 206–223. Leipzig: Teubner, 1898.

Bugaev, Boris. "Shtempelevannaîa kul'tura." *Vesy* 9 (1909), 72–80.

Bugaev, N.V. *Matematika kak orudie nauchnoe i pedagogicheskoe.* Moskva, 1869.

_____ *O svobode voli.* Moskva, 1889.

_____ *Osnovy èvolutsionnoĭ monadologii.* Moskva, 1893.

Butmi, G. *Vragi roda chelovecheskogo.* 4th edn. S.-Peterburg, 1907.

Chamberlain, Houston Stewart. *The Foundations of the Nineteenth Century.* In two vols. Translated by John Lees. 2nd edn. London – New York: The Bodley Head; John Lane Company, 1912.

Chirikov, M.V., and Sheĭnin, O.B. "Perepiska P.A. Nekrasova i K.A. Andreeva." *IMI* 35 (1994), 124–147 (for the English translation, see: http://www.sheynin.de/download/2_Russian%20Papers%20History. pdf, 70–82).

Demidov, S.S. "Iz ranneĭ istorii Moskovskoĭ shkoly teorii funktsiĭ." *IMI* 30 (1986), 124–130.

_____ "N.V. Bugaev i vozniknovenie moskovskoĭ shkoly teorii funktsiĭ deĭstvitel'nogo peremennogo." *IMI* 29 (1985), 113–124.

DOI: 10.1057/9781137338280

Demidov, S.S, and Ford, C.E. "On the Road to a Unified World View: Priest Pavel Florensky – Theologian, Philosopher and Scientist." In *Mathematics and the Divine: A Historical Study.* Edited by T. Koetsier, and L. Bergmans, 595–612. Amsterdam etc.: Elsevier, 2005.

Demidov, S.S., and Levshin, B.V., eds, *Delo akademika Nikolaîa Nikolaevicha Luzina.* Sankt-Peterburg: RKHGI, 1999.

Demidov, S.S, Parshin, A.N., Polovinkin, S.M., and Florenskiĭ, P.V., eds, "Perepiska N.N. Luzina s P.A. Florenskim." *IMI* 31 (1989), 125–191.

Droit, Roger-Pol. *L'Oubli de l'Inde: Une amnésie philosophique.* Paris: Seuil, 2004.

Egorshin, V. *Estestvoznanie, filosofiîa i marksizm.* Moskva: Gosizdat RSFSR, Moskovskiĭ rabochiĭ, 1930.

Èĭkhenbaum, B.M. *Lev Tolstoĭ: issledovaniîa. Stat'i.* Sankt-Peterburg: Fakul'tet filologii i iskusstv SPBGU, 2009.

Filippov, A. *Velikiĭ schet.* Odessa: Vseukrainskoe gosudarstvennoe izdatel'stvo, 1922.

Florenskiĭ, P. "O simvolakh beskonechnosti." *Novyĭ put'* 9 (1904), 173–285.

—— *Stolp i utverzhdenie istiny.* Moskva: Pravda, 1990.

Franz, N., Hagemeister, M., and Haney, F. eds, *Pavel Florenskij – Tradition und Moderne. Beiträge zum Internationalen Symposium an der Universität Potsdam, 5. bis 9. April 2000.* Frankfurt am Main etc.: Peter Lang, 2001.

Ganelin, R.Sh. *Rossiĭskoe samoderzhavie v 1905 g. Reformy i revoliutsiîa.* Sankt-Peterburg: Nauka, S.-Peterburgskoe otdelenie, 1991.

Gladisch, Aug. *Einleitung in das Verständniss der Weltgeschichte. Erste Abtheilung: Die alten Schinesen und die Pythagoreer.* Posen, 1841.

Gopius, E.A. "Filosofiîa 'moskovskoĭ filosofsko-matematicheskoĭ shkoly' i ee otnoshenie k intellektualizmu filosofov XVIII veka i èkonosmicheskomu materializmu K. Marksa." *VFiP* 79: 4 (1905), 554–586.

Graham, Loren, and Kantor, Jean-Michel. *Naming Infinity: A True Story of Religious Mysticism and Mathematical Creativity.* Cambridge, Massachusetts, – London: The Belknap Press of Harvard University Press, 2009.

Guseĭnov, G.Ch. "Politicheskiĭ platonism, obnazhaîushchiĭ i skryvaîushchiĭ." In *Antichnost' i kul'tura Serebrîanogo veka: K 85-letiîu A.A. Takho-Godi,* 533–539. Moskva: Nauka, 2010.

DOI: 10.1057/9781137338280

Hagemeister, Michael, and Kauchtschischwili, Nina, eds, *P.A. Florenskiĭ i kul'tura ego vremeni. P.A. Florenskij e la cultura della sua epoca. Atti del Convegno Internazionale Università degli Studi di Bergamo 10–14 gennaio 1988.* Marburg: Blaue Hörner Verlag, 1995.

Hagemeister, Michael, and Metelka, Torsten, eds, *Appendix 2. Materialien zu Pavel Florenskij.* Berlin u. Zepernick: Kontext, 2001.

Hoefer, Ferdinand. *Histoire des mathématiques: depuis leurs origines jusqu'au commencement du dix-neuvième siècle.* Paris: Librairie Hachette et Cie, 1874.

Izrail' v proshlom, nastoîashchem i budushchem. Sergiev Posad: Izd. "Religiozno-filosofskoĭ Biblioteki," 1915.

ÎUshkevich, A.P. *Istoriîa matematiki v Rossii do 1917 goda.* Moskva: Nauka, 1968.

Kagan, V. "Bugaev Nikolaĭ Vasil'evich." In *Bol'shaîa Sovetskaîa Èntsiklopediîa.* T. 7, 770. Moskva: Gosudarstvennoe slovarno-èntsiklopedicheskoe izd. "Sovetskaîa Èntsiklopediîa," OGIZ RSFSR, 1927.

Kiriânov, ÎU.I., ed., *Pravye partii. 1905–1917. Dokumenty i materialy. V 2 tt.* Moskva: ROSSPÈN, 1998.

Koyré, Alexandre. *La philosophie et le problème national en Russie au début du XIXe siècle.* Paris: Librairie ancienne Honoré Champion, 1929.

Lakhtin, L.K. "Nikolaĭ Vasil'evich Bugaev (Biograficheskiĭ ocherk)." *MS* 25: 2 (1905), 251–269.

Laplace, *Oeuvres complètes*, T. 7. Paris: Gauthier-Villars, Imprimeur-libraire, 1886.

Laruelle, Marlène. *Mythe aryen et rêve impérial dans la Russie du XIXe siècle.* Paris: CNRS, 2005.

Lavrov, P. "Mekhanicheskaîa teoriîa mira." *OZ* 123 (1859), 451–492.

_____ *Ocherk istorii fiziko-matematicheskikh nauk* ([Sankt-Peterburg, 1865–1866]).

Lewes. *Istoriîa filosofii ot nachala ee v Gretsii do nastoîashchikh vremen.* S.-Peterburg, 1865.

Ljunggren, Magnus. *Russkiĭ Mefistofel'. Zhizn' i tvorchestvo Èmiliîa Metnera.* Sankt-Peterburg: Akademicheskiĭ proekt, 2001.

_____ *Twelve Essays on Andrej Belyj's Peterburg.* Göteborg: Göteborgs Universitet, 2009.

Margolin, S.O. "Filosofiîa evreĭskoĭ religii." *Russkaîa mysl'* 3 (1911), 37–41.

Men'shikov, M. "Zvezdy i chisla." *Novoe vremîa*, № 9990, December 25 (January 7), 1903, 7.

DOI: 10.1057/9781137338280

—— "Pis'ma k blizhnim. Vechnoe voskresen'e." *Novoe vremîa*, № 10081, March 28 (April 10), 1904, 3–4.

Mordukhaĭ-Boltovskoĭ, D. "O zakone nepreryvnosti." *VFiP* 87: 2 (1907), 168–184.

Montucla, J. F. *Histoire des mathématiques*. T. 1. Paris: Henri Agasse, 1799.

Na bor'bu za materialisticheskuîu dialektiku v matematike. Moskva – Leningrad: Gosudarstvennoe nauchno-tekhnicheskoe izdatel'stvo, 1931.

Nekrasov, P.A. "Filosofîia i logika nauki o massovykh proîavleniîakh chelovecheskoĭ deîatel'nosti (Peresmotr osnovaniĭ sotsial'noĭ fiziki Quetelet)." *MS* 23: 3 (1902), 463–604.

—— *Gosudarstvo i Akademîia*. Moskva, 1905.

—— "Logika mudrykh lîudeĭ i moral'. (Otvet V.A. Gol'tsevu)." *VFiP* 70: 5 (1903), 902–927.

—— "Moskovskaîa filosofsko-matematicheskaîa shkola i ee osnovateli." *MS* 25: 1 (1904), 3–239.

—— *Teorîia veroîatnosteĭ*. S.-Peterburg, 1912.

—— *Theory of Probability*. Compiled, translated and commented by Oscar Sheynin: http://www.sheynin.de/download/5_Nekrasov.pdf

[Nilus, Sergeĭ], *Bliz grîadushchiĭ Antikhrist i Tsarstvo Diavola na zemle* [Sergiev Posad, 1911].

—— *Na beregu Bozh'eĭ reki. Zapiski pravoslavnogo*. Sergiev Posad, 1916.

Olender, Maurice. *Les Langues du Paradis. Aryens et sémites: un couple providential*. Paris: Seuil, 2002.

Petrova, S.S. "Iz istorii prepodavanîia matematiki v moskovskom universitete s 60-kh gg. XIX – do nachala XX veka." *IMI. Vtoraîa serîia* 46: 11 (2006), 130–147.

Polovinkin, S.M. "O studencheskom matematicheskom kruzhke pri moskovskom matematicheskom obshchestve v 1902–1903 gg." *IMI* 30 (1986), 148–158.

—— "Psikho-aritmo-mekhanik (filosofskie cherty portreta P.A. Nekrasova)." *Voprosy istorii estestvoznanîia i tekhniki* 2 (1994), 109–113.

—— ed., *Sergeĭ Aleksandrovich Nilus (1862–1929)*. Moskva: Izd. Spaso-Preobrazhenskogo Valaamskogo monastyrîa, 1995.

Prasolov, M.A. "'Tsifra poluchaet osobuîu silu' (Sotsial'naîa utopîia Moskovskoĭ filosofsko-matematicheskoĭ shkoly)." *Zhurnal sotsiologii i sotsial'noĭ antropologii* 10: 1 (2007), 38–48.

Sochinenîia Platona. Translated by [V.N.] Karpov. Ch. III. Politika ili Gosudarstvo. Sankt-Peterburg, 1863.

DOI: 10.1057/9781137338280

Pyman, Avril. *Pavel Florensky: A Quiet Genius. The Tragic and Extraordinary Life of Russia's Unknown Da Vinci*. New York – London: Continuum, 2010.

Quételet, Ad. *Du Système social et des lois qui le régissent*. Paris: Guillaumin et Cie, Libraires, 1848.

Repnikov, A.V., and Milevskiĭ, O.A. *Dve zhizni L'va Tikhomirova*. Moskva: Academia, 2011.

Rozanov, V.V. *Sobranie sochineniĭ. Literaturnye izgnanniki. Kniga vtoraîa*. Moskva – Sankt-Peterburg: Respublika; Rostok, 2010.

Schwab, Raymond. *La Renaissance orientale*. Paris: Payot, 1950.

Seneta, Eugene. "Statistical Regularity and Free Will: L.A.J. Quetelet and P.A. Nekrasov." *International Statistical Review* 71: 2 (2003), 319–334.

——"Mathematics, Religion, and Marxism in the Soviet Union in the 1930s." *Historia Mathematica* 31: 3 (2004), 337–367.

Shafarevich, I.R. *Rusofobiîa*. Moskva: Algoritm, 2011.

Shaposhnikov, V.A. "Filosofskie vzglîady N.V. Bugaeva i russkaîa kul'tura kontsa XIX–nachala XX vv." *IMI. Vtoraîa seriîa* 42: 7 (2002), 62–91.

——"K voprosu o filosofsko-metodologicheskikh interesakh N.D. Brashmana." *IMI. Vtoraîa seriîa* 48: 13 (2009), 68–89.

Shestov, A.I., and Alekseev, V.G., eds, *Nauchnaîa pedagogika i russkaîa shkola*. ÎUr'ev, 1916.

Sheynin, Oscar B. "Nekrasov's Work on Probability: The Background." *Archive for History of Exact Sciences* 57 (2003), 337–353.

——"Publikatsii A.A. Markova v gazete 'Den'' za 1914–1915 gg." *IMI* 34 (1993), 194–209 (for the English translation, see: http://www.sheynin.de/download/2_Russian%20Papers%20History.pdf, 63–69).

Solov'ev, S. *Vospominaniîa*. Moskva: Novoe literaturnoe obozrenie, 2003.

Spektorskiĭ, E. *Problema sotsial'noĭ fiziki v XVII stoletii. V 2 tt.* Sankt-Peterburg: Nauka, 2006.

Svetlikova, Ilona. "Kant-semit i Kant-ariets u Belogo." *Novoe literaturnoe obozrenie* 93: 5 (2008), 62–98.

——"Moskovskie pifagoreĭtsy." In S.N. Zenkin, ed., *Intellektual'nyĭ îazyk èpokhi: Istoriîa ideĭ, istoriîa slov*, 117–141. Moskva: Novoe literaturnoe obozrenie, 2011.

Szilard, Lena. *Germetizm i germenevtika*. Sankt-Peterburg: Izd. Ivana Limbakha, 2002.

DOI: 10.1057/9781137338280.

Taube, Mikhail. *Moskovskaîa filosofsko-matematicheskaîa shkola, osnovannaîa prof. Bugaevym i slavîanofil'stvo Khomîakova.* Khar'kov, 1908.

—— *Sovremennyĭ spiritizm i mistitsizm.* Petrograd, 1909.

—— *Svod osnovnykh zakonov myshleniîa.* Petrograd, 1909.

Tokareva, T.A. "Istoriîa matematiki v Rossii: rozhdenie distsipliny," *IMI.* *Vtoraîa seriîa* 44: 9 (2005), 209–237.

Tolstoy, Leo. *War and Peace.* Translated by Louise and Aylmer Maude. Vol.II–III. London: Humphrey Milford, Oxford University Press, 1931–1932.

Vashchenko-Zakharchenko, M.E. *Istoricheskiĭ ocherk matematicheskoĭ literatury indusov.* Kiev, 1882.

Vygodskiĭ, L.S. "Literaturnye zametki ('Peterburg', roman Adreîa Belogo. 1916 g.)," *Novyĭ put',* December 11, 47 (1916), st. 27–32.

Vygodskiĭ, M.ÎA. "Matematika i ee deîateli v Moskovskom universitete vo vtoroĭ polovine XIX v." *IMI* 1 (1948), 141–183.

DOI: 10.1057/9781137338280

Index

DOI: 10.1057/9781137338280

DOI: 10.1057/9781137338280

DOI: 10.1057/9781137338280